SIDEREUS NUNCIUS

SIDEREUS NUNCIUS
or
THE SIDEREAL MESSENGER

GALILEO GALILEI

Translated with
introduction, conclusion, and notes by

ALBERT VAN HELDEN

THE UNIVERSITY OF CHICAGO PRESS
CHICAGO AND LONDON

The University of Chicago Press, Chicago 60637
The University of Chicago Press, Ltd., London

Printed in the United States of America
98 97 96 95 94 93 92 54

Library of Congress Cataloging-in-Publication Data

Galilei, Galileo, 1564–1642.
[Sidereus nuncius. English]
Sidereus nuncius, or, The sidereal messenger / Galileo Galilei; translated with introduction, conclusion, and notes by Albert Van Helden.
p. cm.
Translation of: Sidereus nuncius.
Bibliography: p.
Includes index.
ISBN 0-226-27902-2. ISBN 0-226-27903-0 (pbk.)
1. Astronomy—Early works to 1800. 2. Galilei, Galileo, 1564–1642. Sidereus nuncius. I. Title. II Title: Sidereus nuncius. III. Title: Sidereal messenger.
QB41.G173 1989 88-25179
520—dc19 CIP

∞The paper used in this publication meets the minimum requirements of the American National Standard for Information Sciences—Permanence of Paper for Printed Library Materials, ANSI Z39.48-1984.

CONTENTS

PREFACE

Sidereus Nuncius is not comparable to the great treatises that form the canon of the history of science. It had neither the staying power of Ptolemy's *Almagest* nor the synthetic power of Newton's *Principia*. In fact, as a scientific achievement it cannot be compared to Galileo's own later works, the *Dialogo* and *Discorsi*. There is a good reason for this: *Sidereus Nuncius* was not so much a treatise as an announcement: in a few brief words, and in sober language, it told the learned community that a new age had begun and that the universe and the way in which it was studied would never be the same.

It was an unprecedented sort of book, for although there can be little doubt of Galileo's acute vision and brilliant mind, *Sidereus Nuncius* was the product not of an intellect but rather of an instrument! Heavenly phenomena hidden since the beginning of time were suddenly revealed by the telescope and could be seen by anyone who could procure one of these new devices. The practice, first of astronomy and then of other sciences, was changed. Astronomy ceased to be the exclusive province of learned men: instrument makers without formal training, men of modest talent but great wealth, or autodidacts with manual skill and lots of patience could, and did, henceforth achieve fame in astronomy.[1]

Needless to say, the discoveries announced by Galileo changed the terms of the debate about the world systems. If they did not

1. In the first category we may count the Roman telescope maker Giuseppe Campani (1635–1715), whose telescopes first attracted attention in the 1660s because of his celestial discoveries. In the second, the best example is Johannes Hevelius (1611–87), heir to a brewing fortune in the Polish city of Gdansk, who became well known for his powerful telescopes, his observing zeal, and his sumptuous publications. The best example of the third category is William Herschel (1738–1822), whose tremendously powerful telescopes allowed him to make numerous discoveries and map the Milky Way.

provide logical arguments to tip the balance in favor of the Copernican system, they were nevertheless decisive for they made ancient authority—the foundation of the traditional system of natural philosophy—irrelevent. With *Sidereus Nuncius* we enter the modern world.

Whereas most old scientific treatises do not speak to us easily over the centuries, *Sidereus Nuncius* has lost little of its freshness and we can still share the excitement that gripped those who read it for the first time. It is one of those rare scientific books that is still meaningful and interesting to student and teacher, amateur and scientist alike. For that reason it has been translated from its original Latin into many languages. The first English translation, by Edward Stafford Carlos, was published in 1880 under the title *The Sidereal Messenger of Galileo Galilei and a Part of the Preface to Kepler's Dioptrics.*[2] This work (reprinted in 1960), for many years the standard, is no longer readily available, and its prose is not easily understood by modern students. Stillman Drake's translation in *Discoveries and Opinions of Galileo* (1957) has been the standard English translation for three decades, but it is incomplete. Drake's complete translation in *Telescopes, Tides, and Tactics* (1983) is immersed in a larger narrative. Neither version has the requisite explanatory notes to make Galileo's astronomy accessible to modern students or provide an entry into the detailed literature that has grown up around the book. I have compared my translation with those of Carlos and Drake, with the German translation by Malte Hossenfelder,[3] and with the Italian translation by Maria Timpanaro Cardini.[4]

2. *The Sidereal Messenger of Galileo Galilei and a Part of the Preface to Kepler's Dioptrics containing the original account of Galileo's astronomical discoveries. A translation with introduction and notes by Edward Stafford Carlos* (London, 1880; reprinted, London: Dawsons of Pall Mall, 1960).

3. *Galileo Galilei Sidereus Nuncius Nachricht von neuen Sternen. Dialog über die Weltsysteme (Auswahl). Vermessung der Hölle Dantes. Marginalien zu Tasso. Herausgegeben und eingeleitet von Hans Blumenberg* (Frankfurt am Main: Insel Verlag, 1965), 79–131.

4. *Galileo Galilei Sidereus Nuncius. Traduzione con Testo a Fronte e Note di Maria Timpanaro Cardini* (Florence: Sansoni, 1948).

The present translation is based on the original Latin text in the Venice 1610 edition of *Sidereus Nuncius*.[5] This text is followed with only a few deviations in volume 3 of *Le Opere di Galileo Galilei*. But in the few instances where the texts differed, I have followed the original. I have tried to produce a text which is accurate and yet accessible to today's students. The most difficult problem was that of the title. Should the Latin word *nuncius* be rendered in English as *messenger* or *message?* As Galileo was preparing his book for the printer, he referred to it in his correspondence as his *avviso astronomico,* or *astronomical message* (see p. 20, below), and we may assume that this meaning was his original intent. The request to the Council of Ten (see p. 34, below) for permission to have it printed (dated 26 February 1610) referred to the book as *Astronomica Denuntiatio ad Astrologos,* or *Astronomical Announcement to Students of the Heavens*.[6] When printing was started, the text was introduced by the title *Astronomicus Nuncius* (see p. 35, below). But Galileo changed his mind again, and the title page, printed last, bears the more ambiguous wording *Sidereus Nuncius*. Galileo's contemporaries, most notably Johannes Kepler, interpreted *nuncius* as *messenger,* as can be seen from Kepler's *Dissertatio cum Nuncio Sidereo* or *Conversation with the Sidereal Messenger* (see pp. 94–99, below). This latter interpretation was followed in the first French translation, printed in 1681.[7]

In the English-speaking world, Ormsby MacKnight Mitchel,

5. I have used the facsimile of the copy in the Biblioteca Nazionale Centrale of Florence (Pal 1200/23), of which 1000 copies were printed in 1964, on the occasion of the four-hundredth year since Galileo's birth. In this copy the overlay which substitutes *Medicea* for *Cosmica* on p. 5 (see p. 19, below) is missing. A facsimile lacking the plates was published by Éditions culture et civilisation (Brussels) in the 1960s. A new facsimile was published in 1987 by Archival Facsimiles, Ltd. (Alburgh, Harleston, Norfolk, U.K.).

6. *Opere* 19:227–28. I have followed here Edward Rosen's translation of *astrologos* as *students of the heavens*. See "The Title of Galileo's *Sidereus Nuncius*," *Isis* 14 (1950): 289. Unless otherwise noted, all translations of documents in the *Opere* are my own.

7. *Le messager céleste,* tr. Alexandre Tinelis, Abbé de Castelet (Paris, 1681). An excellent modern French translation is available: *Sidereus Nuncius; le message céleste. Texte établi et trad. par Émile Namer* (Paris: Gauthier-Villars, 1964).

founder of the Cincinnati Observatory, published a popular astronomical journal entitled *Sidereal Messenger* from 1846 to 1848,[8] and four decades later W. W. Payne, director of the Carleton College Observatory in Northfield, Minnesota, used the same title for his journal that was the forerunner of George Ellery Hale's *Astrophysical Journal*.[9] In the meantime the first English translation of *Sidereus Nuncius* was published in London by Edward Stafford Carlos, who likewise chose *Sidereal Messenger* as his title.[10] In 1950 Edward Rosen reviewed the history of this "error" and showed that Galileo himself, in 1626, argued against the "messenger" interpretation.[11] After Rosen's article, there could be little doubt as to Galileo's original intent.[12] But the problem persisted.

An entire generation of English-speaking students of the history of science has benefited from Stillman Drake's incomplete translation of *Sidereus Nuncius*, contained in *Discoveries and Opinions of Galileo*. Drake was, of course, well aware that Galileo originally meant *nuncius* in the sense of *message*. Yet he translated the title as *Starry Messenger* for reasons of tradition, while rendering *astronomicus nuncius* in the heading of the first page of the text (see p. 35, below) as *astronomical message*.[13] When criticized for this by Rosen

8. *The Sidereal Messenger, a monthly journal devoted to astronomical science*, ed. O. M. Mitchel (Cincinnati, 1846–48).

9. *The Sidereal Messenger*, ed. W. W. Payne, 10 vols. (Northfield, MN, 1882–91). Starting with vol. 11 (1892), the title was *Astronomy and Astro-physics*, and George Ellery Hale became coeditor. In 1894 Hale became the sole editor of the journal and changed its name to *Astrophysical Journal* (vols. 1–; Chicago, 1894–).

10. See note 2.

11. Edward Rosen, "The Title of Galileo's *Sidereus Nuncius*," *Isis* 41 (1950): 287–89.

12. Note that in modern translations into other languages Galileo's original intent is followed. *Nuncius* was translated by Maria Timpanaro Cardini into Italian as *annunzio*, by Émile Namer into French as *message*, and by Malte Hossenfelder into German as *Nachricht*. José Fernandes Chitt, however, translated it into Spanish as *mensajero*. See *El Mensajero de los Astros*, tr. J. Fernandes Chitt, intr. José Babini (Buenos Aires: Editorial Universitaria de Buenos Aires, 1964).

13. *Discoveries and Opinions of Galileo, translated with an introduction and notes by Stillman Drake* (Garden City, NY: Doubleday & Co., 1957), 19, 27. Note that in *Telescopes, Tides, and Tactics* (1983) Drake maintains these translations (pp. 12, 17).

in a review,[14] Drake defended his choice at some length. He pointed out that in 1610 Galileo's own student, among others, used the "messenger" interpretation (see p. 101, below) and that for over a decade Galileo himself did not object to that interpretation, by which time it had become firmly established. In other words Galileo allowed the "error" to take root, and perhaps it pleased rather than irritated him. Furthermore, Drake argued, it was perfectly consistent to consider the book the messenger and its content the message.[15]

On balance I agree with Drake. Although it is abundantly clear from his correspondence in the first few months of 1610 that Galileo meant *message, messenger* has the sanction of tradition, a tradition that Galileo allowed to grow by his initial silence. In keeping with tradition in the English-speaking world, then, I have chosen *Sidereal Messenger* as the subtitle of this book, and following Drake, I have translated *astronomicus nuncius* at the head of the text as *astronomical message*.

In preparing this work for publication, I have received help from several colleagues. Helen Eaker checked my initial translation and saved me from a number of errors. Stillman Drake, Owen Gingerich, and Noel Swerdlow read the entire work and made many helpful comments. Robert O'Dell helped me with several astronomical problems, and George Trail suggested many ways in which my prose could be improved. The redrawing of several of Galileo's diagrams was done by Philip Sadler. The title page of the first

14. Edward Rosen, "Stillman Drake's *Discoveries and Opinions of Galileo,*" *Isis* 48 (1957): 440–43.

15. Stillman Drake, "The Starry Messenger," *Isis* 49 (1958): 346–47. In *Telescopes, Tides, and Tactics,* a hypothetical dialogue, Drake put the following words in the mouth of Sarpi (p. 12): "The title he [Galileo] had had in mind was 'Astronomical Message,' as will be seen on the first page of the text. But it then occurred to him that what carries a message is a messenger and that a very attractive title for a book containing news from the stars would be the one chosen. Thus it was the book, and not its author, that is there called *nuncius,* or ambassador—though the same word may also mean simply *message.*"

edition of *Sidereus Nuncius* as well as the original illustrations used in the translation are reproduced from the copy at Wellesley College.[16] I am grateful to Wellesley College for allowing me to reproduce these, and I thank Katharine Park and Anne Anninger for helping me obtain them. Among my students in History 223 at Rice University in the fall of 1987 who patiently read an imperfect early draft, I must single out Philip Samms for his comments. I remain, of course, responsible for any errors. Rice University supported my research generously in too many ways to enumerate. Part of this work was carried out under a grant from the National Science Foundation.

16. The first edition of the *Sidereus* omitted one of the "Medicean stars" from four figures (those reproduced on pages 72, 74, 78, and 81 of this book), even though these stars were shown in Galileo's manuscript. I have added these missing stars in square brackets.

INTRODUCTION

In the autumn of 1609 Galileo Galilei, a forty-five-year-old professor of mathematics at the University of Padua (near Venice), directed a twenty-powered telescope at the Moon, setting off a chain of events that was to shake the intellectual edifice of Europe to its foundations. Although by no means the first scientist to use the new device in this way, Galileo was by far the most successful. He made the first crucial discoveries with the instrument and dominated this new area of inquiry as no one has since.

When Galileo first heard about it, in the summer of 1609, the telescope was very new. Reports of a strange new device that showed faraway things as though nearby had begun spreading from the Netherlands the previous autumn. But before pursuing those events, let us take a look at the background.

Our story begins with spectacle lenses. Up to about 1300, aging scholars suffering from what we now know to be a progressive inability to focus their eyes on nearby things were seriously hampered in their reading and writing, usually beginning in their middle forties.[1] This condition, called *presbyopia,* effectively limited the careers of many men of letters. A solution came in the late thirteenth century. In his *Opus Maius* of 1267, the Franciscan friar Roger Bacon (ca. 1214–92) wrote about magnifying glasses, thick sections of glass spheres that could be laid over one's reading material in order

1. In the eye, light is refracted by the cornea, the aqueous humor, the lens, and the vitreous body, but only the lens can accommodate. When the lens is flatter, in its unaccommodated shape, light from distant objects is focused on the retina; when it is more convex, in its accommodated shape, light from nearby objects is focused on the retina. The flexibility of the lens decreases progressively, and in their forties most people begin experiencing difficulties in focusing on things closer than about 2 feet. At this point reading becomes difficult and reading glasses, simple convex lenses, must be used to supply the extra refractive power needed for close work. See Jane F. Koretz and George H. Handelman, "How the Human Eye Focuses," *Scientific American* 259, no. 1 (July 1988): 92–99.

to make the letters easier to read. He mentioned that such glasses were useful to the aged, "for they can see a letter, no matter how small, sufficiently enlarged."[2] Bacon speculated on the powers of technology (for which he was considered a magician) and made some extravagant claims about the miraculous effects that could be achieved by means of glasses.[3]

Before the end of the thirteenth century craftsmen in Italy had begun making thin, biconvex glasses and putting them into frames so that they could be worn in front of the eyes.[4] These glasses, thicker in the middle than at the edges, were shaped like lentils (*lens* in Latin), hence our word "lens." From this point forward, the elderly could avail themselves of reading glasses, although we must not think that these early spectacles were very comfortable or that the quality of the lenses was very high.

By the middle of the fifteenth century spectacle makers in Italy were also making concave glasses for aiding "the weak vision of the young," that is *myopia*.[5] It appears that these early versions could only correct mild degrees of myopia because it was difficult to grind and polish highly curved concave glasses. But both concave and convex lenses were now in circulation, and as the spectacle-making craft (usually organized in a guild) spread from Italy to other parts of Europe, young and old could enjoy the benefits of these wonderful devices. And not only the major cities were served: itinerant peddlers hawked their optical wares throughout the countryside's small settlements, markets, and fairs.

2. Roger Bacon, *Opus Maius,* tr. Robert B. Burke, 2 vols. (Philadelphia: University of Pennsylvania Press, 1928), 2:574.

3. Ibid., p. 582.

4. Edward Rosen, "The Invention of Eyeglasses," *Journal for the History of Medicine and Allied Sciences* 11 (1956): 13–46, 183–218.

5. Vincent Ilardi, "Eyeglasses and Concave Lenses in Fifteenth-Century Florence and Milan: New Documents," *Renaissance Quarterly* 29 (1976): 341–60. In the myopic eye the cornea and lens combined have too much refractive power so that light rays from distant objects come to a focus in front of the retina. A myope can therefore not see distant objects sharply but can bring objects very close to the eye into focus.

If concave and convex lenses were available all over Europe by about 1500, why didn't the telescope appear at this time? After all, a telescope can be made by combining a convex and a concave, or two convex lenses. The answer to this question can be found in the strengths of the lenses. In order to achieve a noticeable and useful magnifying effect one must combine a weak convex lens with a strong concave or convex lens. It appears that the appropriate range of strengths simply was not available at this stage, nor would it be until the turn of the seventeenth century.

In the meantime, as "natural magic" blossomed in the sixteenth century, speculation developed about the wonderful effects that might be achieved by means of lenses and mirrors. For this reason it has sometimes been argued that some early form of telescope may very well have been in use during this period.[6] But the claims of some sixteenth-century *magi* that optical devices with miraculous powers could be made were never translated into actual telescopes or microscopes because they were not based on a proper understanding of the optical principles involved. It seems clear, however, that in Italy combinations of lenses were being used in some applications (such as efforts to correct defective vision) by the end of the sixteenth century, and that the telescope was very much "in the air." Moreover, during this period Italian glassworkers were exporting their know-how to other parts of Europe, including the Netherlands.[7]

The age of the telescope began late in September 1608, in the Netherlands. On 25 September members of the provincial government of Zeeland, the southwestern province of the newly formed Dutch Republic, wrote to their representative at the national government, the States-General in The Hague, that a spectacle maker in the city of Middelburg (the capital of Zeeland) had "a certain device by means of which all things at a very great distance can be

6. For an examination of this argument, see Albert Van Helden, *The Invention of the Telescope.* American Philosophical Society, *Transactions* 67, part 4 (1977): 5–16.

7. Ibid., p. 24.

seen as if they were nearby."[8] A few days later the States-General discussed the patent application of Hans Lipperhey on such a device. Within two weeks, however, two other claimants for the invention came forward, Jacob Metius of Alkmaar (north of Amsterdam) and Sacharias Janssen of Middelburg. The States-General decided that the invention, although very useful, should not receive a patent because it was too easy to copy.[9] It appears that at about the same time that Lipperhey was in The Hague requesting his patent, a Dutch peddler was offering the same device for sale at the annual autumn fair at Frankfurt, three hundred miles southeast of The Hague.[10]

This device was obviously no secret. Within weeks of Lipperhey's patent application, the news was spreading out from Holland through diplomatic channels. Since the spyglass was so easy to copy, the news of its existence was followed quickly by the instrument itself. By the spring of 1609 little spyglasses were offered for sale by spectacle makers in Paris, and by the summer the device had reached Italy.[11] These gadgets, consisting of a convex and a concave lens in a tube, magnified only three or four times, and therefore some of the men whose interest had been piqued by the rumors about their miraculous effect were disappointed when they actually examined one.[12]

One man who was not disappointed was Galileo. The rumor of the new device first reached his friend, the theologian Paolo Sarpi, in Venice in November 1608.[13] The following spring, Father Sarpi

8. Ibid., pp. 20, 35–36.

9. Ibid., pp. 20–25, 35–44.

10. Ibid., pp. 21–23.

11. Ibid., pp. 25–28.

12. Ibid., pp. 44–45.

13. Paolo Sarpi to Francesco Castrino, 9 December 1608, in Manlio Duilio Busnelli, "Un Carteggio Inedito di Fra Paolo Sarpi con l'Ugonotto Francesco Castrino," *Atti del Reale Istituto Veneto di Scienze, Lettere ed Arti* 87, part 2 (1927–28): 1069; reprinted in *Fra Paolo Sarpi, Lettere ai Protestanti,* ed. Busnelli, 2 vols. (Bari: Gius. Laterza & Figli, 1931), 2:15. See also Sarpi to Jerome Groslot de l'Isle, 9 January 1609, ibid., 1:58.

wrote to Paris to ask Jacques Badovere, a former student of Galileo's, for confirmation.[14] It is at this point that Galileo first concerned himself with the new instrument. In *Sidereus Nuncius* he tells us that he first heard the rumor around May 1609 and that it was confirmed shortly afterward by Badovere's letter (p. 37, below). We may speculate that perhaps he had heard the rumor earlier but had paid little attention to it (extravagant claims for mysterious devices that turned out to be hoaxes were common). But when he heard from Sarpi that Badovere had confirmed the existence of the device and probably reported that spyglasses were commonly for sale in Paris, he became very interested. It was easy for Galileo to obtain spectacle lenses and reproduce the invention. Indeed, he stated later that he had done so on the first night after he returned from Venice (where he had presumably been shown Badovere's letter by Sarpi).[15] In this effort Galileo was not exceptional: several others had already done as much. What he did with the new device over the next 6 months was, however, crucial to the history of astronomy.

A Tuscan by birth (born in Pisa and brought up in Florence), Galileo had since 1592 taught mathematics at the University of Padua in the Venetian Republic. As an oldest son, he had heavy financial responsibilities (such as providing dowries for his sisters), and although he was not married, he had two daughters and a son by his mistress. In order to supplement his salary, which was inadequate to his financial needs, he took in student boarders and sold scientific instruments made by a craftsman in his employ. In

14. Sarpi to Badovere, 30 March 1609. See Busnelli, "Un Carteggio Inedito," 1160. Sarpi received letters from Paris 3 to 5 weeks after they were written. We may assume therefore that Badovere's reply could not have reached him much before the middle of May. This fits well with Galileo's statement (p. 36, below) that he heard the rumor and that it was confirmed to him "about 10 months ago," which, counting back from the middle of March 1610, would place that event in the middle of May 1609.

15. Galileo, *The Assayer* (1623). See Stillman Drake and C. D. O'Malley, *The Controversy on the Comets of 1618* (Philadelphia: University of Pennsylvania Press, 1960), 211.

the midst of all this, he carried on his researches on motion. By 1609, when the events recounted here overtook him, he had arrived at a number of important conclusions, including the epoch-making law of falling bodies. Like many professors, however, he was always alert to opportunities to improve his financial position and to gain more time for research. He now seized on the spyglass.

Galileo's first effort, put together from ordinary spectacle lenses, magnified three times (p. 37, below), and he immediately set himself the task of making the instrument more powerful. As an experienced teacher of mathematical subjects, he was familiar with optical theory, but the science of optics of his day could tell him little about the workings of the spyglass. He was a brilliant experimenter, however, and through trial and error he quickly figured out that the magnification of this simple instrument depends on the ratio of the focal lengths of the two lenses. Having determined this relationship, Galileo knew that he needed weaker convex lenses and/or stronger concave lenses. The problem was that such lenses could not be bought from spectacle makers because these artisans made lenses only in a narrow range of strengths. Galileo therefore had to teach himself to grind and polish the required lenses, an arduous task requiring considerable manual skill. Toward the end of August 1609 he had managed to construct a spyglass that magnified eight or nine times (p. 37, below). This was an enormous improvement over the common spyglasses that were now coming into Venice.[16] Through the offices of his good friend Paolo Sarpi, Galileo now approached the Senate of Venice, wishing to demonstrate his new instrument. In a letter written on 29 August he relates what happened next:[17]

> . . . it is 6 days since I was called by the doge to whom I had to show it together with the entire Senate, to the infinite

16. *Opere,* 10:250, 255.

17. Ibid., p. 253. I have followed, with minor alterations, the translation in Stillman Drake, *Galileo at Work: His Scientific Biography* (Chicago: University of Chicago Press, 1978), p. 141. See also Edward Rosen, "The Authenticity of Galileo's Letter to Landucci," *Modern Language Quarterly* 12 (1951): 473–86.

amazement of all; and there have been numerous gentlemen and senators who, though old, have more than once climbed the stairs of the highest bell towers of Venice to observe at sea sails and vessels so far away that, coming under full sail to port, 2 hours and more were required before they could be seen without my spyglass. For in fact the effect of this instrument is to represent an object that is for example 50 miles away as large and near as if it were but 5.

The gentlemen were greatly impressed by the obvious military advantage of this instrument. Two days later Galileo appeared before the Senate and donated his instrument to the republic. He accompanied the gift with a letter to the doge, the chief magistrate of Venice, written in the style then customary when addressing a ruler:[18]

Most Serene Prince,

Galileo Galilei, a most humble servant of Your Serene Highness, being diligently attentive, with all his spirit, not only to discharging the duties pertaining to the lecturing of mathematics at the University of Padua, but also to bringing extraordinary benefit to Your Serene Highness with some useful and remarkable invention, appears now before You with a new contrivance of glasses [*occhiale*], drawn from the most recondite speculations of perspective, which renders visible objects so close to the eye and represents them so distinctly that those that are distant, for example, 9 miles appear as though they were only 1 mile distant. This is a thing of inestimable benefit for all transactions and undertakings, maritime or terrestrial, allowing us at sea to discover at a much greater distance than usual the hulls and sails of the enemy, so that for 2 hours and more we can detect him before he detects us and, distinguishing the number and kind of the vessels, judge his force, in order to prepare for chase, combat, or flight; and likewise, allowing us on land to look inside the

18. *Opere,* 10:250–51.

> fortresses, billets, and defenses of the enemy from some prominence, although far away, or also in open campaign to see and to distinguish in detail, to our very great advantage, all his movements and preparations; besides many other benefits, clearly manifest to all judicious persons. And therefore, judging it worthy to be received by Your Serene Highness and esteemed very useful, I have decided to present it to You and to submit the decision about this invention to Your power, so that you may ordain and provide, according as it seems opportune to Your providence, whether or not it will be built.
>
> And with all affection the said Galileo presents this to Your Serene Highness as one of the fruits of the science which he has professed for the past 17 years at the University of Padua, with the hope of carrying on his work in order to present to You greater ones, if it shall please the Good Lord and Your Serene Highness that he, according to his desire, will pass the rest of his life in Your service. For which he bows down humbly, and from His Divine Majesty he prays for the utmost of all happiness for You.

In other words Galileo gives the doge and the Senate sole rights to the manufacture of his instrument, and in return he asks very tactfully for some improvement of his position at the university. After his presentation Galileo was told that his contract at the university would be renewed for life (in other words he was given tenure) and that his salary would be increased from its current 480 to 1000 florins per year.[19] (When he received official notification, however, he learned to his disappointment that the new salary would not go into effect until the expiration of his current contract, at the end of the 1609–10 school year, and that the award excluded further salary increases.)[20]

As mentioned above, Galileo was not the only one in the Venetian Republic who had a spyglass. Travelers from other regions were

19. Ibid., p. 254.
20. Ibid., 19:116–17.

coming into Venice, offering to sell for considerable sums simple spyglasses that magnified three or four times. By devoting all his ingenuity and energy to improving the device, Galileo had managed to gain a substantial lead on the competition with his more powerful instrument. If, in the midst of this feverish activity, he took the time to point this device to the heavens, he was not the first to do so. One of the first spyglasses had already been pointed to the stars in Holland in the autumn of 1608,[21] and a few weeks before Galileo presented his instrument to the doge, Thomas Harriot in England had turned a six-powered spyglass on the Moon and had drawn the first telescopic likeness of that body—not much better than what could be achieved with the naked eye.[22] At this time Galileo was obviously more concerned with the rewards to be reaped from the earthly advantages of an improved instrument than with any celestial advantage.

Perhaps driven by his ambition to improve his situation even further, hopefully at the court of his native Tuscany this time, Galileo continued his efforts to make better and more powerful spyglasses. Sometime that autumn he began studying the heavens, especially the Moon, through these instruments, and we may assume that at this time he began to realize that the instrument would revolutionize astronomy and cosmology. When, sometime in November 1609, he finished an instrument that magnified twenty times—more than twice as much as his August effort—he undertook his first astronomical research project, a thorough study of the Moon. Between 30 November and 18 December he observed and drew our satellite as it went through its phases, leaving no fewer than eight drawings.[23]

21. *Ambassades du Roy de Siam envoyé à l'Excellence du Prince Maurice, arrivé à la Haye le 10. Septemb. 1608* (The Hague, 1608), p. 11. See Stillman Drake, *The Unsung Journalist and the Origin of the Telescope* (Los Angeles: Zeitlin & Ver Brugge, 1976).

22. Terrie Bloom, "Borrowed Perceptions: Harriot's Maps of the Moon," *Journal for the History of Astronomy* 9 (1978): 117–22.

23. The question as to the dates of Galileo's lunar observations was examined by Guglielmo Righini in "New Light on Galileo's Lunar Observations," *Reason,*

What intrigued Galileo, and others after him, about the Moon was the irregularity of its surface as revealed by the new instrument. According to the then prevailing geocentric cosmology of Aristotle, the heavens were perfect and unchanging, and heavenly bodies were perfectly smooth and spherical. The large spots visible on the Moon

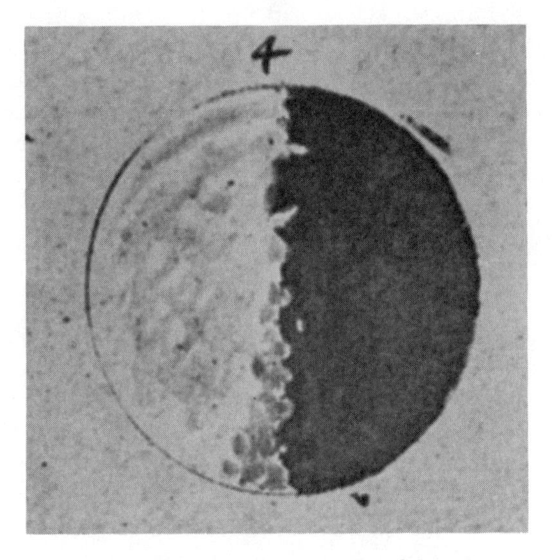

One of Galileo's wash drawings of the Moon
(*Le Opere di Galileo Galilei* 3 (1892): 48).

Experiment, and Mysticism in the Scientific Revolution, ed. Maria Luisa Righini Bonelli and William Shea (New York: Science History Publications, 1975), 59–76. See also Owen Gingerich, "Dissertatio cum Professore Righini at Sidereo Nuncio," ibid., pp. 77–88; Stillman Drake, "Galileo's First Telescopic Observations," *Journal for the History of Astronomy* 7 (1976): 153–68, at 153–54; and Righini, *Contributo alla Interpretazione Scientifica dell'Opera Astronomica di Galileo,* monograph 2, *Annali dell'Istituto e Museo di Storia della Scienza* (Florence, 1978), pp. 26–44. I have followed the conclusions of Ewen A. Whitaker in "Galileo's Lunar Observations and the Dating of the Composition of 'Sidereus Nuncius,' " *Journal for the History of Astronomy* 9 (1978): 155–69.

to the naked eye were usually explained away by ad hoc devices. One could, for instance, postulate that parts of the perfectly smooth Moon absorbed and then emitted light differently from other parts.[24] But the Copernican theory had, so to speak, put the Earth in the heavens and thus tended to blur the distinction between the earthly realm of change and corruption and the unchanging and perfect heavens. Furthermore, the new star of 1572 (a supernova) and the demonstration that the comet of 1577 was in the heavens and not (as Aristotelians contended) in the earthly region had dealt heavy blows to the immutability and perfection of the heavens. Yet, few had fully assimilated the implications of these new developments into their conceptual apparatus. When Galileo examined the Moon with his twenty-powered spyglass, its surface appeared anything but smooth: it seemed rough and uneven. The dividing line between light and darkness (the terminator) was not a smoothly curved line at all, as one would expect if the Moon's surface was perfectly smooth. Instead it was very irregular. In the bright part features were starkly outlined by black lines that grew broader and narrower as the light of the Sun varied; in the dark part there were little patches of light. Galileo drew the conclusion that the Moon's surface is full of mountains, valleys, and plains, just as the Earth's surface is. In a letter dated 7 January 1610, the first letter describing telescopic observations, he put it this way:[25]

> . . . it is seen that the Moon is most evidently not at all of an even, smooth, and regular surface, as a great many people believe of it and of the other heavenly bodies, but on the contrary it is rough and unequal. In short it is shown to be such that sane reasoning cannot conclude otherwise than that it is full of prominences and cavities similar, but much larger, to the mountains and valleys spread over the Earth's surface.

24. Roger Ariew, "Galileo's Lunar Observations in the Context of Medieval Lunar Theory," *Studies in the History and Philosophy of Science* 15 (1984): 213–26.

25. *Opere,* 10:273. I follow the translation by Drake in "Galileo's First Telescopic Observations," *Journal for the History of Astronomy* 7 (1976): 153–68, at 155.

Galileo went on to describe the phenomena in considerable detail, rehearsing, as it were, the observations and conclusions he was to publish more elaborately a few months later in *Sidereus Nuncius*. He also mentioned that he was about to finish an instrument that would enlarge things thirty times, adding: "Of all the above observations, none is seen or can be seen without an exquisite instrument, because of which we can believe to have been the first in the world to discover something about the heavenly bodies from so nearby and so distinctly."[26]

He was thus fully aware by this time of the historical importance of what he was observing. What he did not know yet, on 7 January, was that another observation he mentioned would catapult him to fame in the next three months:[27]

> And besides the observations of the Moon I have observed the following in the other stars. First, that many fixed stars are seen with the spyglass that are not discerned without it; and only this evening I have seen Jupiter accompanied by three fixed stars, totally invisible because of their smallness; and the configuration was in this form:

East * * * * West

As mentioned above, others had already noticed that with a spyglass one can see many more stars than with the naked eye. But Galileo here drew the configuration of what he thought were three fixed stars near the planet Jupiter. His attention had been drawn to them by the fact that they formed a straight line with Jupiter and

26. *Opere,* 10:277; Drake, "Galileo's First Telescopic Observations," 157. I have made minor changes in Drake's translation.

27. *Opere,* 10:277; Drake, "Galileo's First Telescopic Observations," 157, with minor changes in the translation.

that they were very bright for their size. In the following days he was to discover that they were not fixed stars at all.

The letter also mentions that through the spyglass the planets appear in the form of little globes, like little moons, while the fixed stars do not—again a point on which he was to elaborate in *Sidereus Nuncius*. Lest we think that all these observations were easy, Galileo cautioned his correspondent:[28]

> . . . the instrument must be held firm, and hence it is good, in order to escape the shaking of the hand that arises from the motion of the arteries and from respiration itself, to fix the tube in some stable place. The glasses should be kept very clear and clean by means of a cloth, or else they become fogged by the breath, humid or foggy air, or by the vapor itself which evaporates from the eye, especially when it is warm.

It should be noted that a telescope of the type used by Galileo, one with a concave ocular that shows objects right-side up, has a very small field of view. At magnifications of twenty or more, such an instrument will show at most half the diameter of the Moon. This small field makes the instrument difficult to use, especially when not rigidly mounted. Small celestial objects such as Jupiter are by no means easy to find with it and extremely difficult to keep in the field.

This letter of 7 January is the first scientific discussion of telescopic phenomena, and, except for its mention of the satellites of Jupiter, it is an outline of *Sidereus Nuncius,* published 9 weeks later. But it was his discovery of those bodies that started the chain of events that made Galileo famous.

Galileo's spyglasses of up to twenty magnifications were adequate for observing earthly objects and the Moon, but when it came to very small and bright objects such as the fixed stars and planets, their optical imperfections severely limited their usefulness. Even spherical lenses with perfectly uniform curvature suffer from

28. *Opere,* 10:277–78; Drake, "Galileo's First Telescopic Observations," 158.

the defects of spherical and chromatic aberration[29] (defects about which Galileo knew nothing), but in addition these early lenses were of a very uneven curvature. As a result, images, especially of objects that are very bright in proportion to their size, such as candle flames or stars, were ill defined and surrounded by prismatic colors.

In his efforts to improve the images for purposes of observing the stars and planets, Galileo hit on the device of stopping down the aperture of his telescopes by means of putting cardboard rings in front of the objective lens, thus limiting the incident light to the area near the optical axis where the curvature of the lenses was more uniform. In his letter of 7 January he wrote: "It is good that the convex glass, which is the one far from the eye, should be partly covered and that the opening left should be oval in shape, since thus are objects seen much more distinctly."[30] The oval shape of the aperture ring must mean that in this particular instrument the objective lens was astigmatically ground, an indication of just how primitive these lenses were. Surviving instruments of this period have their apertures stopped down drastically, so that a twenty-powered spyglass typically had an aperture of 1.5 to 2.5 cm, that is, it only multiplied the light-gathering power of the eye a few times. Since the focal lengths of these objectives were of the order of one meter, these spyglasses had a focal ratio of f/50 or more, and this was to remain the rule for the rest of the seventeenth century. No doubt, Galileo was the first to adapt a spyglass in this manner, and we may assume that this happened shortly before he wrote the letter of 7 January.

29. In a lens with spherical curvature, incident rays parallel to the optical axis will not all unite at the same point. This defect, known as spherical aberration, was known to several writers at the time. Although it is possible that Galileo was familiar with it, there is no mention of it in his writing at this time. Furthermore, in its passage through the lens incident light is broken up into the colors of the spectrum, and the different colors come to a focus at different distances. This defect, known as chromatic aberration, was first pointed out by Isaac Newton in 1672. Both defects are more noticeable in the light that passes through the outer parts of the lens.

30. *Opere,* 10:278; Drake, "Galileo's First Telescopic Observations," 158.

Having thus improved the image quality of small bright objects, Galileo began a program of observing the planets and fixed stars. He could now see that the planets could be resolved into little globes or disks, while the stars, or "fixed stars" as they were then called,[31] showed no disks. This was important confirmation of the Copernican hypothesis, to which Galileo had subscribed quietly for some time,[32] for according to Copernicus the fixed stars were much farther away than the planets.

Not all planets were in good positions for observation. In the beginning of January 1610 Venus was in the morning sky, while Saturn and Mars were close to the Sun and near their greatest distances from the Earth. Galileo was able to notice little about these planets except that they showed disks. Things were different in the case of Jupiter. This planet had just passed opposition (when it is on a straight line with the Earth and Sun and at its closest approach to the Earth) and was the brightest object in the evening sky. When Galileo turned his newly adapted spyglass to it on the evening of 7 January, his attention was drawn to the formation shown earlier. He thought, of course, he was seeing three little fixed stars in a row and that Jupiter just happened to be passing through their formation that evening. Near opposition, Jupiter's motion with respect to the fixed stars is retrograde, that is, from east to west,[33] and therefore, when Galileo again sought out Jupiter the next evening, he expected to see the stars in the same formation,

31. In the cosmology of Aristotle, heavenly bodies were absolutely different from earthly objects. But all heavenly bodies were made of the same celestial substance. All heavenly bodies were therefore called stars. The vast majority, the "fixed stars," rotated about the Earth in unchanging formation. They provided the background against which the motions of the seven "wandering stars" were plotted as they moved through the zodiac. Our word *planet* derives from the Greek word for wanderer.

32. Stillman Drake, "Galileo's Steps to Full Copernicanism and Back," *Studies in the History and Philosophy of Science* 18 (1987): 93–105.

33. All planets, including the Earth, move about the Sun in the same direction, from west to east, the inner ones faster than the outer ones. When a planet beyond the Earth is at opposition, the faster motion of the Earth makes it appear as though the planet is moving backward through the zodiac, that is, from east to west.

with Jupiter having moved to the west with respect to them. What he saw in fact was that Jupiter had moved to the *east*, still on the same straight line. This puzzled him, and he thought that perhaps the astronomical tables were wrong and Jupiter had returned to its direct, west to east, motion. Jupiter's seemingly anomalous behavior greatly intrigued him.

On the ninth of January it was cloudy, but on the tenth he could again observe the planet. To his surprise he now saw only two of the stars and Jupiter was to the *west* of both of them. Obviously the planet's motion was retrograde after all. But how could the appearances change in this manner? Over the next week of observations he found that there were in fact *four* little stars, that Jupiter was not moving away from them but remained always on the straight line near them, and that the stars moved along the straight line with respect to each other and Jupiter. By 15 January at the latest he had the solution to this strange behavior: Jupiter had four moons!

The novelty and importance of this discovery can hardly be overstated. Since time immemorial there had been only seven wanderers, the Moon and Sun, Mercury, Venus, Mars, Jupiter, and Saturn. Now, all of a sudden, one of them was shown to have four companions, wanderers completely unknown to the great philosophers of Antiquity. And they answered a major criticism against the Copernican theory: if the Earth were a planet, why should it be the only one to have a moon going around it, and how could there be two centers of motion in the universe? Now it was evident that the Earth was not the only planet to have a moon, and that no matter what world system one subscribed to, there was more than one center of motion. And if Jupiter's moons were more glamorous, the philosophical implications of the uneven lunar surface were even greater. Here was objective evidence that heavenly bodies were not perfect.

How long would it be before others discovered the same phenomena? Galileo knew that he must publish his discoveries as quickly as possible. The roughness of the Moon's surface was within the grasp of instruments less powerful than his, given a good observer. Moreover, the moons of Jupiter formed a bright and arresting formation around the planet, apparent to anyone who had managed to improve the spyglass to magnifications comparable to his instrument.[34] Obviously, time was of the essence. Galileo did not want to be scooped, the more so because he had not given up hope of improving his position in life. The Venetian Senate offered little hope for further advancement. Galileo had over the years kept his contacts with his native Tuscany very much alive. A few years earlier, in the summer of 1605, he had acted as mathematics tutor to young Cosimo de' Medici, who in 1609 became Grand Duke Cosimo II of Tuscany, and he kept in frequent touch with the Medici court. These new and dazzling discoveries offered Galileo an opportunity to obtain patronage from the ruler of his native land.

While continuing his observations of Jupiter's moons and drawing several examples of the myriads of telescopic stars visible in known constellations and asterisms, Galileo now wrote up his discoveries. Earlier, presumably during a quick visit to Florence in the autumn,[35] he had shown the Grand Duke what the Moon looked like through one of his early spyglasses. He now wrote a brief report of his discoveries (dated 30 January) to the Tuscan court:[36]

> I am at present in Venice to have printed some observations concerning the heavenly bodies which I have made with one of my spyglasses, and since they are of infinite amazement, I infinitely render grace to God that it has pleased him to make me alone the first observer of an admirable thing, kept hidden

34. Near opposition, the four Galilean satellites of Jupiter have stellar magnitudes between 5 and 6. Were it not for the planet's brightness, they would just be visible with the naked eye. There is evidence that at least one of them was observed with the naked eye in China. See Xi Ze-zong, "The Sighting of Jupiter's Satellite by Gan De 2000 Years before Galileo," *Chinese Astronomy and Astrophysics* 5 (1981): 242–43; and David W. Hughes, "Was Galileo 2,000 Years Too Late?" *Nature* 296 (18 March 1982): 199.

35. Stillman Drake, *Galileo at Work,* 142.

36. *Opere,* 10:280–81.

all these ages. That the Moon is a body very similar to the Earth was already ascertained by me and shown in part to Our Most Serene Lord, although imperfectly since I did not yet have a spyglass of the excellence that I have now. Besides the Moon, this spyglass has allowed me to discover a multitude of fixed stars never before seen, of which there are more than ten times as many as are naturally visible. Moreover, I have assured myself about what has always been a controversy among the philosophers, that is, what is the Milky Way. But, what exceeds all wonders, I have discovered four new planets and observed their proper and particular motions, different among themselves and from the motions of all the other stars; and these new planets move about another very large star[37] like Venus and Mercury,[38] and perchance the other known planets, move about the Sun. As soon as this tract, which I shall send to all the philosophers and mathematicians as an announcement, is finished, I shall send a copy to the Most Serene Grand Duke, together with an excellent spyglass, so that he can verify all these truths.

Galileo quickly heard that Grand Duke Cosimo and his three brothers were "astonished by this new proof of [his] almost supernatural intelligence,"[39] and now he made a very shrewd move. On 13 February he wrote to the Grand Duke's secretary as follows:[40]

As to my new observations, I shall indeed send them as an announcement to all philosophers and mathematicians, but not without the favor of our Most Serene Lord. For since God graced me with being able, through such a singular sign, to reveal to my Lord my devotion and the desire I have that his

37. Jupiter.

38. In a variation of the Ptolemaic system, Venus and Mercury moved about the Sun.

39. *Opere,* 10:281.

40. *Ibid.,* p. 283.

> glorious name live as equal among the stars, and since it is up to me, the first discoverer, to name these new planets, I wish, in imitation of the ancient sages who placed the most excellent heroes of that age among the stars, to inscribe these with the name of the Most Serene Grand Duke. There only remains in me a little indecision whether I should dedicate all four to the Grand Duke only, calling them *Cosmian* [*Cosmica*][41] after his name or, rather, since they are exactly four in number, dedicate them to all four brothers with the name *Medicean Stars*.

By return mail the secretary notified him that the latter choice was the preferred one.[42] Galileo had thought, however, that the Grand Duke would prefer *Cosmian Stars* (or *Cosmic Stars*), and the printing had already begun by the time the secretary's response arrived. On the first page of the text, the new planets were therefore referred to as *Cosmica Sydera*. This error was corrected by pasting slips of paper with the word *Medicea* over *Cosmica* in most if not all copies of the book (see p. 35, below).[43] Galileo continued his observations of the new planets as the printing progressed. His last observation is dated 2 March 1610. At the last moment he also decided to expand the section on the fixed stars by adding actual illustrations (and explanatory text) of some star formations. The four pages containing this material were added unnumbered in the middle of the book. It appears that Galileo was still putting the finishing touches on the manuscript as the printing was nearing completion, for the last few pages of the surviving manuscript are

41. The Grand Duke's name, Cosimo, was latinized as *Cosmus*. *Cosmica* is the adjectival form of *Cosmus* as well as *cosmos,* the Greek word for universe or world. *Cosmica* could mean "Cosmian," "cosmic," or "wordly." This confusion led Belisario Vinta to prefer *Medicean*. In a forthcoming article entitled "Galileo the Emblem-Maker," Mario Biagioli documents the importance of the planet Jupiter in the mythology of the Medici family.

42. *Opere,* 10:284–85.

43. Antonio Favaro, *Galileo Galilei e lo Studio di Padova,* 2 vols. (Padua, 1883); 2d ed., Padua: Antenore, 1966), 1:299–300.

full of additions and corrections, and appear to be little more than a first draft.[44]

The dedicatory letter of *Sidereus Nuncius* is dated 12 March 1610, and on the next day Galileo sent an advance, unbound copy, accompanied by a letter, to the Tuscan court.[45] On 19 March he sent a properly bound copy together with the instrument with which he had made the discoveries, so that the Grand Duke would be able to see the new celestial phenomena for himself. He mentioned also that the 550 copies that had been printed had already been sold and announced his plan for a second, enlarged edition.[46] But nothing ever came of this plan.

Galileo chose the title *Sidereus Nuncius* for his little book after some deliberation. As explained in the Preface, the word *nuncius* can mean both *messenger* and *message*. Because he used the Italian word *avviso,* that is, *announcement* or *dispatch,* to refer to the book in his correspondence (and even the title *Avviso Astronomico*),[47] we may assume he meant the word in the latter sense, and we should therefore translate the title as *Starry Message* or *Sidereal Message*. But many of Galileo's contemporaries, including Johannes Kepler, took the meaning of *nuncius* to be *messenger,* and Galileo did not object to this interpretation for many years. The tradition of referring to the work as the *Starry* or *Sidereal Messenger* therefore took root, and I have not departed from it.

The book opens with a flowery dedicatory letter in which Galileo praises the noble Cosimo II and dedicates the new planets to him. Such letters were common practice until well into the nineteenth century, as long as funding for science was heavily dependent on personal patronage. Galileo was writing for two audiences, his

44. *Opere,* 3, part 1: 46–47. Note also that on the last two pages of the book the print is squeezed and full of abbreviations. See *Sidereus Nuncius* (Venice, 1610), f. 28.

45. *Opere,* 10:288–89.

46. *Ibid.,* p. 300.

47. *Ibid.,* pp. 283, 288, 298, 300.

prospective patron Cosimo II and his scientific peers. As we shall see, this approach bore fruit.

The main body of *Sidereal Messenger* begins with a standard rhetorical introduction in which the excellence and novelty of the discoveries are briefly adduced. There follows a concise description of the instrument and how it functions, and only then is Galileo ready to discuss his discoveries. This discussion consists of two lengthy sections, one about the Moon and one about Jupiter's satellites, separated by a brief discussion of the fixed stars and the difference in appearance between stars and planets. There is very little in the way of conclusion, which is not surprising in view of the hurried nature of the undertaking.

The section on the Moon is, as it were, a little treatise in itself. It represents Galileo's first telescopic researches, and it is therefore not surprising that for coherence and cogency of argument it is the best part of the book. The reader is treated to a description of the motion of light and shadow over the face of the Moon, a motion revealing its craggy nature. It is a compelling verbal portrait of our satellite, supported by visual evidence. The quality of the engravings leaves something to be desired, but their effect is nevertheless profound. Throughout his description of the lunar phenomena, Galileo makes comparisons with earthly phenomena, thus forcefully stressing the Moon's Earth-like nature. At one point he goes so far as to compare a large, round central valley (in all likelihood the crater Albategnius) with Bohemia, a large earthly plain surrounded by mountains. Moreover, the affinity between the two bodies is explicitly tied to the Pythagorean view of the universe, which was then frequently used (although erroneously) to refer to the Copernican theory.

Knowing full well the revolutionary nature of these discoveries, Galileo anticipated arguments. If the Moon has mountains, then why does not the edge appear like the outline of a toothed wheel? He correctly explains that the valleys at the edge of the visible hemisphere are filled in by the peaks of successive ranges of mountains in front of and behind them. Thus the imperfections in the circular outline are largely obliterated. In fact, it would be another five decades before telescopes were good enough to show the remaining small irregularities in the Moon's outline.[48] As another partial explanation, Galileo put forward the possibility that, like the Earth, the Moon was surrounded by an envelope of a substance denser than the ether (although he later withdrew this explanation).

How high are the mountains on the Moon? In a nice piece of geometry, Galileo derived a figure from the lengths of the shadows. His conclusion was that their height was greater than four miles, which to his mind made them higher than earthly mountains.

The dark part of the Moon does not always appear completely dark. For some time before and after the new Moon (i.e., conjunction with the Sun), the dark part appears illuminated by an ashen light. Over the centuries several explanations had been proposed for this phenomenon, but the current explanation could not be found as long as thinkers were convinced that the Earth is a base body that is dark and does not reflect light. Galileo gave the correct explanation. When the Moon is new, the Earth as seen from the Moon is full. Light from the Sun is reflected by the Earth and illuminates the Moon, thus making its dark part, held in the grip of the thin sickle of bright light, faintly visible from the Earth. He was not the first to give this explanation: it had first been suggested by Leonardo da Vinci, in an unpublished notebook, a century earlier;[49] Johannes Kepler relates that his teacher Michael Maestlin, a Copernican, had already published it in a set of theses printed in 1596, now lost;[50] and Kepler himself had published a full explanation in his *Astronomia pars Optica* of 1604.[51]

48. These were first seen by Giovanni Domenico Cassini in 1664. See Giuseppe Campani, *Ragguaglio di due Nuove Osservazioni* (Rome, 1664), 38–40.

49. Leonardo Da Vinci (1452–1519), Codex Leicester-Hammer, f. 2r. See Jane Roberts, *Le Codex Hammer de Léonard de Vinci, les eaux, la terre, l'univers* (Paris: Jacquemart-André, 1982), 12, 30.

50. Edward Rosen, *Kepler's Conversation with Galileo's Sidereal Messenger* (New York: Johnson Reprint Corp., 1965), 32, 117–19.

51. *Ad Vitellionem Parilapomena in quo Astronomia Pars Optica Traditur* (1604), *Johannes Kepler Gesammelte Werke,* 2:223–24.

Galileo next turned his attention to the stars and planets. First he pointed out that the stars are not enlarged by the telescope as the Moon is. Their brightness is increased, but their size is only slightly enlarged. The planets, on the other hand, were resolved by the telescope into clearly outlined round disks, like little moons. It is here, in the observation of these small bodies, very bright compared to their size, that Galileo's lead in telescope technology is most evident. But we must not allow this instrumental superiority to obscure Galileo's uncanny talent as an observer. He grasped the fact that the magnified stellar images presented by his telescope were at least in large part spurious. His successors, talented observers themselves who used much better telescopes, sometimes still believed that they were observing resolved stellar disks of measurable diameter.

There was thus a great difference in size between stars and planets, and the logical conclusion was that the stars, although often very bright, are much farther away from us than the planets. This supported the Copernican theory, in which, because of the absence of annual stellar parallax, the fixed stars had been moved very far away from the Sun and Earth, leaving a huge gap between Saturn, believed to be the outermost planet, and the sphere of the fixed stars. But Galileo did not mention the Copernican theory in this context.

This section continues with a description of the innumerable fixed stars visible through the telescope. Galileo showed two examples, the region of the belt and sword of Orion, and the Pleiades, both well-known formations. He had planned to map the entire constellation of Orion, but the great number of stars in it had forced him to select a smaller region. The section ends with a discussion of the Milky Way and other nebular patches. Galileo argued that these nebulosities were resolved into aggregates of many small stars by the telescope and showed as examples the nebulous areas in the head of Orion and Praesepe, the Beehive (in Antiquity called the Manger), in Cancer.

The most lengthy portion of the book deals with the new planets discovered about Jupiter. Here Galileo presents a straightforward factual account, discussing all his observations from 7 January to 2 March. Had he merely described his discovery and given one or two examples of the formation of Jupiter and its companions, his claim would have been less convincing. The lengthy sequence of observations familiarized the reader with the motions of the satellites, as well as the motion of the entire formation with respect to the fixed stars, and showed the care with which the observations had been made. Galileo summarized his results as follows: four moons orbit Jupiter, while Jupiter orbits the center of the world; these moons travel about Jupiter in orbits of different sizes, and the smaller the orbit the shorter the period. Although Galileo had not yet been able to determine the periods, he indicated that the period of the closest moon was of the order of a day or so, whereas that of the farthest was about half a month.

Here Galileo took the opportunity to strike a blow for the Copernican system, although he stopped short of declaring his allegiance to that world system. In the Ptolemaic system the Earth was the single center of all celestial motions; in the Copernican system there were two centers of motion, the Sun and the Earth. Why, opponents of the Copernican system asked, should the Earth be the only planet to have a moon? The telescope supplied the answer: the Earth is not the only planet with a moon; Jupiter has no fewer than four.

There remained a small problem. The apparent sizes of Jupiter's moons appeared to Galileo to vary from time to time. He explained this by postulating that Jupiter, like the Earth and Moon, had an envelope of matter denser than the rest of the ether, which partially obscured the satellites when interposed between them and the eye of the observer.

Here, as he had done in the section on the Moon, Galileo promised the reader a more ample treatment in a forthcoming work on "the system of the world." That book was not to appear until 1632. In the meantime he ended the tract with a promise that the reader might soon expect more from his pen.

SIDEREUS NUNCIUS

SIDEREAL MESSENGER

unfolding great and very wonderful sights
and displaying to the gaze of everyone,
but especially philosophers and astronomers,
the things that were observed by

GALILEO GALILEI,

Florentine patrician[1]
and public mathematician of the University of Padua,
with the help of a spyglass[2] lately devised[3] by him,
about the face of the Moon, countless fixed stars,
the Milky Way, nebulous stars,
but especially about
four planets
flying around the star of Jupiter at unequal intervals
and periods with wonderful swiftness;
which, unknown by anyone until this day,
the first author detected recently
and decided to name

MEDICEAN STARS[4]

1. Galileo came from a Florentine family that can be traced back to the thirteenth century. His ancestors included several members of the governing council of the Florentine Republic and a celebrated physician. His family tree can be found in *Opere,* 19:17. See also Stillman Drake, *Galileo at Work,* 448.

2. The Latin word used here is *perspicillum.* Galileo used the Italian word *occhiale* to describe the instrument. I have translated these terms as *spyglass* throughout. The word *telescope* was unveiled only in 1611. See p. 112, below.

3. Galileo used the Latin word *reperti,* from the verb *reperio.* This word can mean both *invented* and *devised.* Although Galileo was often accused of claiming he actually invented (in our sense) the telescope, this is clearly a calumny, as demonstrated by the passage on pp. 36–37, below. See Edward Rosen, "Did Galileo Claim He Invented the Telescope?" *Proceedings of the American Philosophical Society* 98 (1954): 304–12.

4. Galileo referred to Jupiter's satellites as both "planets" and "stars." In the old terminology, based on Aristotelian cosmology, both terms were correct. See also note 31, p. 15.

SIDEREVS NVNCIVS

MAGNA, LONGEQVE ADMIRABILIA
Spectacula pandens, suspiciendaque proponens
vnicuique, præsertim verò
PHILOSOPHIS, atq; ASTRONOMIS, quæ à

GALILEO GALILEO
PATRITIO FLORENTINO

Patauini Gymnasij Publico Mathematico

PERSPICILLI

Nuper à se reperti beneficio sunt obseruata in LVNÆ FACIE, FIXIS INNVMERIS, LACTEO CIRCVLO, STELLIS NEBVLOSIS,
Apprime verò in

QVATVOR PLANETIS

Circa IOVIS Stellam disparibus interuallis, atque periodis, celeritate mirabili circumuolutis; quos, nemini in hanc vsque diem cognitos, nouissimè Author depræhendit primus; atque

MEDICEA SIDERA

NVNCVPANDOS DECREVIT.

VENETIIS, Apud Thomam Baglionum. M DC X.

Superiorum Permissu, & Priuilegio.

MOST SERENE
COSIMO II DE' MEDICI
FOURTH GRAND DUKE OF TUSCANY[5]

A most excellent and kind service has been performed by those who defend from envy the great deeds of excellent men and have taken it upon themselves to preserve from oblivion and ruin names deserving of immortality. Because of this, images sculpted in marble or cast in bronze are passed down for the memory of posterity; because of this, statues, pedestrian as well as equestrian, are erected; because of this, too, the cost of columns and pyramids, as the poet says,[6] rises to the stars; and because of this, finally, cities are built distinguished by the names of those who grateful posterity thought should be commended to eternity. For such is the condition of the human mind that unless continuously struck by images of things rushing into it from the outside, all memories easily escape from it.

Others, however, looking to more permanent and long-lasting things, have entrusted the eternal celebration of the greatest men not to marbles and metals but rather to the care of the Muses and to incorruptible monuments of letters. But why do I mention these things as though human ingenuity, content with these [earthly]

5. Cosimo II de' Medici (1590–1621) was the grandson of Cosimo I, the first of the family to bear the title of Grand Duke. He ascended the throne in 1609 upon the death of his father, Ferdinand I.

6. The reference is to the *Elegies* of the Roman poet Sextus Propertius, who lived in the last half of the first century B.C. Book 3, no. 2, is on the power of song and reads in part: "For not the heaven-raised Pyramids' expense, / Nor Jove's house which, at Ellis, mimics heaven, / Nor Mausulus, his tomb's magnificence, / By Death's supreme indemnity forgiven. / To filching fire or rain their crowns submit, / By Time's stroke, and their weight, they crash, defied. / Not so shall pass the fame by poet's wit / Achieved; for that endures in deathless pride." See E. H. W. Meyerstein, *The Elegies of Propertius* (London: Oxford University Press, 1935), 95–96.

realms, has not dared to proceed beyond them? Indeed, looking far ahead, and knowing full well that all human monuments perish in the end through violence, weather, or old age, this human ingenuity contrived more incorruptible symbols against which voracious time and envious old age can lay no claim. And thus, moving to the heavens, it assigned to the familiar and eternal orbs of the most brilliant stars the names of those who, because of their illustrious and almost divine exploits, were judged worthy to enjoy with the stars an eternal life. As a result, the fame of Jupiter, Mars, Mercury, Hercules, and other heroes by whose names the stars are addressed will not be obscured before the splendor of the stars themselves is extinguished. This especially noble and admirable invention of human sagacity, however, has been out of use for many generations, with the pristine heroes occupying those bright places and keeping them as though by right. In vain Augustus's affection tried to place Julius Caesar in their number, for when he wished to name a star (one of those the Greeks call *Cometa* and we call hairy)[7] that had appeared in his time the Julian star, it mocked the hope of so much desire by disappearing shortly.[8] But now, Most Serene Prince, we are able to augur truer and more felicitous things for Your Highness, for scarcely have the immortal graces of your

7. Both the Greek *cometes* and Latin *crinitus* mean *hairy.* The original meaning was thus *hairy star,* describing the appearance of these celestial objects.

8. In an English translation of Suetonius's biographies of the first twelve caesars made during Galileo's lifetime, we read, in the 88th section of the life of Julius Caesar: "He died in the 56 yeare of his age and was canonized among the Gods, not onely by their voice who decreed such honour unto him, but also by the perswasion of the common people. For at those Games and playes which were the first that Augustus his heire exhibited for him thus deified, there shone a blazing starre for seven dayes together, arising about the eleventh houre of the daye; and beleeved it was to be the soule of Caesar received up into heaven. For this cause also uppon his Image there is a starre set to the verie Crowne of his head." See *Suetonius History of Twelve Caesars translated into English by Philemon Holland anno 1606,* 2 vols. (London: David Nutt, 1899), 1:80. See also Wilhelm Gundel and Hans Georg Gundel, *Astrologumena: Die Astrologische Literatur in der Antike und ihre Geschichte,* beiheft 6, *Sudhoffs Archiv* (Wiesbaden: Franz Steiner, 1966), 127–28.

soul begun to shine forth on earth than bright stars offer themselves in the heavens which, like tongues, will speak of and celebrate your most excellent virtues for all time. Behold, therefore, four stars reserved for your illustrious name, and not of the common sort and multitude of the less notable fixed stars, but of the illustrious order of wandering stars, which, indeed, make their journeys and orbits with a marvelous speed around the star of Jupiter, the most noble of them all, with mutually different motions, like children of the same family, while meanwhile all together, in mutual harmony, complete their great revolutions every twelve years about the center of the world, that is, about the Sun itself.[9] Indeed, it appears that the Maker of the Stars himself, by clear arguments, admonished me to call these new planets by the illustrious name of Your Highness before all others. For as these stars, like the offspring worthy of Jupiter, never depart from his[10] side except for the smallest distance, so who does not know the clemency, the gentleness of spirit, the agreeableness of manners, the splendor of the royal blood, the majesty in actions, and the breadth of authority and rule over others, all of which qualities find a domicile and exaltation for themselves in Your Highness? Who, I say, does not know that all these emanate from the most benign star of Jupiter, after God the source of all good? It was Jupiter, I say, who at Your Highness's birth, having already passed through the murky vapors of the horizon, and occupying the midheaven[11] and illuminating the eastern angle[12] from his royal house, looked down upon Your most fortunate birth from that sublime throne and poured out all his splendor and grandeur into the most pure air, so that with its

9. Clearly Galileo is referring here to the Copernican system.

10. While in recent times it has become customary in the English language to refer to heavenly bodies with the personal pronoun *it,* until the nineteenth century the Sun, Mercury, Mars, Jupiter, and Saturn were referred to as *he* and the Moon and Venus as *she.*

11. The midheaven is the intersection of the ecliptic and the meridian.

12. This is the *horoscopus,* the point of the ecliptic rising at the eastern horizon marking the beginning of the first house.

first breath Your tender little body and Your soul, already decorated by God with noble ornaments, could drink in this universal power and authority. But why do I use probable arguments when I can deduce and demonstrate it from all but necessary reason? It pleased Almighty God that I was deemed not unworthy by Your serene parents to undertake the task of instructing Your Highness in the mathematical disciplines, which task I fulfilled during the past four years, at that time of the year when it is the custom to rest from more severe studies. Therefore, since I was evidently influenced by divine inspiration to serve Your Highness and to receive from so close the rays of your incredible clemency and kindness, is it any wonder that my soul was so inflamed that day and night it reflected on almost nothing else than how I, most desirous of Your glory (since I am not only by desire but also by origin and nature under Your dominion), might show how very grateful I am toward You. And hence, since under Your auspices, Most Serene Cosimo, I discovered these stars unknown to all previous astronomers, I decided by the highest right to adorn them with the very august name of Your family. For since I first discovered them, who will deny me the right if I also assign them a name and call them the Medicean Stars,[13] hoping that perhaps as much honor will be added to these stars by this appellation as was brought to other stars by the other heroes? For, to be silent about Your Most Serene Highness's ancestors to whose eternal glory the monuments of all histories testify,[14] Your virtue alone, Great Hero, can, by Your name, impart immortality to these stars. Indeed, who can doubt that You will not only meet but also surpass by a great margin the highest expectation

13. The telescope inaugurated a new chapter in celestial discovery. By claiming the right to name his discoveries, Galileo set a trend that others were to follow, with varying degrees of success, into the twentieth century. Systems of naming celestial objects are now regulated by international agreement, and names are often assigned by a committee of the International Astronomical Union.

14. For a history of the Medici family, see Ferdinand Schevill, *The Medici* (New York: Harcourt, Brace & Co., 1949; New York: Harper, 1960); and J. R. Hale, *Florence and the Medici: The Pattern of Control* (London: Thames & Hudson, 1977).

raised by the most happy beginning of your reign, so that when You have surpassed Your peers You will still contend with Yourself, which self and greatness You are daily surpassing.

Therefore, Most Merciful Prince, acknowledge this particular glory reserved for You by the stars and enjoy for a very long time these divine blessings carried down to You not so much from the stars as from the Maker and Ruler of Stars, God.

Written in Padua on the fourth day before the Ides of March,[15] 1610.

Your Highness's most loyal servant,

Galileo Galilei

15. In formal letters such as this one, writers often used the Roman manner of designating days of the month, in which days were counted backward from the kalends, nones, or ides, beginning with the day of the kalends, nones, or ides itself. The ides occurred on the fifteenth day of March, May, July, and October, and on the thirteenth day of all other months. The fourth day before the Ides of March is therefore 12 March.

The undersigned Gentlemen, Heads of the Council of Ten,[16] having received certification from the Reformers of the University of Padua,[17] by report from the Gentlemen deputized for this matter, that is, from the Most Reverend Father Inquisitor and from circumspect Secretary of the Senate, Giovanni Maraviglia, with an oath, that in the book entitled *Sidereus Nuncius* by Galileo Galilei there is nothing contrary to the Holy Catholic Faith, Principles, or good customs, and that it is worthy of being printed, allow it a license so that it can be printed in this city.

Written on the first day of March 1610

M. Ant. Valaresso
Nicolo Bon — Heads of the Council of Ten
Lunardo Marcello

The Secretary of the Most Illustrious Council of Ten

Bartholomaeus Cominus

1610, on 8 March. Registered in the book on p. 39

Ioan. Baptista Breatto
Coadjutor of the Congregation on Blasphemy

16. The Council of Ten, first instituted in 1310 as a committee of public safety and made a permanent institution in 1335, dealt with all criminal and moral matters. It also exercised power in foreign affairs, finance, and war. Its heads granted permission to print books.

17. The *Riformatori dello Studio di Padova* constituted the body of overseers of the university. Since 1517 it had been made up of three members of the Venetian Senate. The *riformatori* were charged by the government with censorship of the press in the Venetian territories. They made recommendations to the Council of Ten. See Paul F. Grendler, "The Roman Inquisition and the Venetian Press, 1540–1605," *Journal of Modern History* 47 (1975): 48–65; reprinted in *Culture and Censorship in Late Renaissance Italy and France* (London: Variorum Reprints, 1981), no. 9.

ASTRONOMICAL MESSAGE

Containing and Explaining Observations Recently Made, With the Benefit of a New Spyglass, About the Face of the Moon, the Milky Way, and Nebulous Stars, about Innumerable Fixed Stars and also Four Planets hitherto never seen, and named MEDICEAN STARS

In this short treatise I propose great things for inspection and contemplation by every explorer of Nature. Great, I say, because of the excellence of the things themselves, because of their newness, unheard of through the ages, and also because of the instrument with the benefit of which they make themselves manifest to our sight.

Certainly it is a great thing to add to the countless multitude of fixed stars visible hitherto by natural means and expose to our eyes innumerable others never seen before, which exceed tenfold the number of old and known ones.[18]

It is most beautiful and pleasing to the eye to look upon the lunar body, distant from us about sixty terrestrial diameters,[19] from so near as if it were distant by only two of these measures, so that the diameter of the same Moon appears as if it were thirty times, the surface nine-hundred times, and the solid body about twenty-seven thousand times larger than when observed only with the

18. In the star catalog in his *Almagest,* Ptolemy listed 1022 stars. See G. J. Toomer, *Ptolemy's Almagest* (London: Duckworth, 1984), 341–99.

19. The distance of the Moon was commonly known to be about sixty terrestrial *radii.* In the manuscript as well as the printed version of *Sidereus Nuncius,* Galileo mistakenly uses *diameters,* as he does in his letter of 7 January 1610 (*Opere,* 10:273, 277). A slip of the pen therefore appears to be ruled out. See Edward Rosen, "Galileo on the Distance between the Earth and the Moon," *Isis* 43 (1952): 344–48.

naked eye.[20] Anyone will then understand with the certainty of the senses that the Moon is by no means endowed with a smooth and polished surface, but is rough and uneven and, just as the face of the Earth itself, crowded everywhere with vast prominences, deep chasms, and convolutions.

Moreover, it seems of no small importance to have put an end to the debate about the Galaxy or Milky Way and to have made manifest its essence to the senses as well as the intellect; and it will be pleasing and most glorious to demonstrate clearly that the substance of those stars called nebulous up to now by all astronomers is very different from what has hitherto been thought.

But what greatly exceeds all admiration, and what especially impelled us to give notice to all astronomers and philosophers, is this, that we have discovered four wandering stars, known or observed by no one before us. These, like Venus and Mercury around the Sun,[21] have their periods around a certain star[22] notable among the number of known ones, and now precede, now follow, him, never digressing from him beyond certain limits. All these things were discovered and observed a few days ago by means of a glass contrived by me after I had been inspired by divine grace.

Perhaps more excellent things will be discovered in time, either by me or by others, with the help of a similar instrument, the form and construction of which, and the occasion of whose invention, I shall first mention briefly, and then I shall review the history of the observations made by me.

About 10 months ago a rumor came to our ears that a spyglass

20. Galileo implies here that in these observations he used an instrument that magnified thirty times. In his letter of 7 January 1610, he stated that he was about to finish a thirty-powered instrument (*Opere,* 10:277), but there is no evidence that he made much use of this instrument. See Drake, *Galileo at Work,* 147–48.

21. In the traditional Ptolemaic scheme, all planets were thought to orbit the Earth. In a well-known variation of this scheme that may well have been suggested in Greek Antiquity, Mercury and Venus were thought to orbit the Sun. This explained the fact that they never stray far from the Sun.

22. See p. 15, note 31, above.

had been made by a certain Dutchman[23] by means of which visible objects, although far removed from the eye of the observer, were distinctly perceived as though nearby. About this truly wonderful effect some accounts were spread abroad, to which some gave credence while others denied them. The rumor was confirmed to me a few days later by a letter from Paris from the noble Frenchman Jacques Badovere.[24] This finally caused me to apply myself totally to investigating the principles and figuring out the means by which I might arrive at the invention of a similar instrument, which I achieved shortly afterward on the basis of the science of refraction.[25] And first I prepared a lead tube in whose ends I fitted two glasses,[26] both plane on one side while the other side of one was spherically convex and of the other concave. Then, applying my eye to the concave glass, I saw objects satisfactorily large and close. Indeed, they appeared three times closer and nine times larger than when observed with natural vision only.[27] Afterward I made another more perfect one for myself that showed objects more than sixty times larger.[28] Finally, sparing no labor or expense, I progressed so far that I constructed for myself an instrument so excellent that things seen through it appear about a thousand times larger and more than

23. The Latin word *Belga* should be translated as *Dutchman* or *Netherlander.* See "A Note on the Word 'Belgium,' " in Pieter Geyl, *The Netherlands in the Seventeenth Century, Part I, 1609–1648* (London: Ernest Benn, 1961), 260–62.

24. See pp. 4–5, above.

25. As a professor of mathematical subjects, Galileo was thoroughly grounded in the optical theory of his day. This theory could not, however, give him much guidance in duplicating the invention. In *The Assayer* of 1623, Galileo more fully described the process by which he figured out how to make his first spyglass. See Stillman Drake and C. D. O'Malley, *The Controversy on the Comets of 1618* (Philadelphia: University of Pennsylvania Press, 1960), 211–13.

26. The Latin word *perspicillum* was here clearly meant to denote a common spectacle lens.

27. This was the greatest magnification that could be achieved with a spyglass made with lenses for sale in the shops of spectacle makers.

28. This is the instrument Galileo presented to the Venetian Senate. See pp. 6–8, above.

thirty times closer than when observed with the natural faculty only.[29] It would be entirely superfluous to enumerate how many and how great the advantages of this instrument are on land and at sea. But having dismissed earthly things, I applied myself to explorations of the heavens. And first I looked at the Moon from so close that it was scarcely two terrestrial diameters[30] distant. Next, with incredible delight I frequently observed the stars, fixed as well as wandering,[31] and as I saw their huge number I began to think of, and at last discovered, a method whereby I could measure the distances between them. In this matter, it behooves all those who wish to make such observations to be forewarned. For it is necessary first that they prepare a most accurate glass that shows objects brightly, distinctly, and not veiled by any obscurity, and second that it multiply them at least four hundred times and show them twenty times closer. For if it is not an instrument such as that, one will try in vain to see all the things observed in the heavens by us and enumerated below. Indeed, in order that anyone may, with little trouble, make himself more certain about the magnification of the instrument, let him draw two circles or two squares on paper, one of which is four hundred times larger than the other, which will be the case when the larger diameter is twenty times the length of the other diameter. He will then observe from afar both sheets fixed to the same wall, the smaller one with one eye applied to the glass and the larger one with the other, naked eye. This can easily be done with both eyes open at the same time. Both figures will then appear of the same size if the instrument multiplies objects according to the desired proportion. After such an instrument has been prepared, the method of measuring distances is to be investigated, which is achieved by the following procedure. For the sake of easy comprehension, let *ABCD* be the tube and *E* the eye of the observer. When there are no glasses in the tube, the rays proceed

29. See note 17.
30. See note 16.
31. That is, stars and planets.

to the object *FG* along the straight lines *ECF* and *EDG,* but with the glasses put in they proceed along the refracted lines *ECH* and *EDI.* They are indeed squeezed together and where before, free, they were directed to the object *FG,* now they only grasp the part *HI.* Then, having found the ratio of the distance *EH* to the line *HI,* the size of the angle subtended at the eye by the object *HI* is found from the table of sines, and we will find this angle to contain only some minutes, and if over the glass *CD* we fit plates perforated some with larger and some with smaller holes, putting now this plate and now that one over it as needed, we form at will angles subtending more or fewer minutes. By this means we can conveniently measure the spaces between stars separated from each other by several minutes with an error of less than one or two minutes.[32] Let it suffice for the present, however, to have touched on this so lightly and to have, so to speak, tasted it only with our lips, for on another occasion we shall publish a complete theory of this instrument.[33] Now let us review the observations made by us during the past 2 months, inviting all lovers of true philosophy to the start of truly great contemplation.

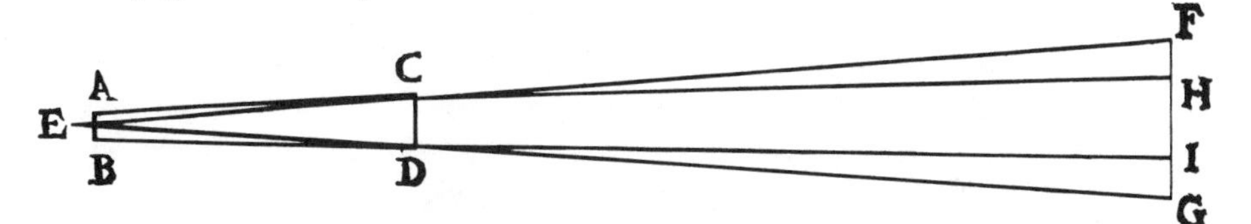

Let us speak first about the face of the Moon that is turned toward our sight, which, for the sake of easy understanding, I divide into two parts, namely a brighter one and a darker one. The brighter

32. The relationship between the size of the aperture of the objective lens and the field of view of the instrument is, in fact, rather more complicated than Galileo implies here, and for this reason all efforts to turn this form of telescope into a measuring instrument failed. See John North, "Thomas Harriot and the First Telescopic Observations of Sunspots," in John W. Shirley, ed., *Thomas Harriot; Renaissance Scientist* (Oxford: Clarendon Press, 1974), 129–65, at 158–60.

33. Galileo never published such a theory.

part appears to surround and pervade the entire hemisphere, but the darker part, like some cloud, stains its very face and renders it spotted. Indeed, these darkish and rather large spots are obvious to everyone, and every age has seen them. For this reason we shall call them the large or ancient spots, in contrast with other spots, smaller in size and occurring with such frequency that they besprinkle the entire lunar surface, but especially the brighter part. These were, in fact, observed by no one before us.[34] By oft-repeated observations of them we have been led to the conclusion that we certainly see the surface of the Moon to be not smooth, even, and perfectly spherical, as the great crowd of philosophers have believed about this and other heavenly bodies,[35] but, on the contrary, to be uneven, rough, and crowded with depressions and bulges. And it is like the face of the Earth itself, which is marked here and there with chains of mountains and depths of valleys. The observations from which this is inferred are as follows.

On the fourth or fifth day after conjunction,[36] when the Moon displays herself to us with brilliant horns,[37] the boundary dividing the bright from the dark part does not form a uniformly oval line, as would happen in a perfectly spherical solid, but is marked by an uneven, rough, and very sinuous line, as the figure shows. For several, as it were, bright excrescences extend beyond the border between light and darkness into the dark part, and on the other hand little dark parts enter into the light. Indeed, a great number of small darkish spots, entirely separated from the dark part, are distributed everywhere over almost the entire region already bathed by the light of the Sun, except, at any rate, for that part affected

34. On Thomas Harriot's telescopic observation of the Moon in August 1609, see p. 9, above.

35. See pp. 10–11, above.

36. That is, conjunction with the Sun, when the Moon is invisible because its illuminated hemisphere is turned away from the Earth. The current astronomical term is "new moon." It is at this point that solar eclipses can occur.

37. That is, the Moon shows only a thin crescent of light.

by the large and ancient spots. We noticed, moreover, that all these small spots just mentioned always agree in this, that they have a dark part on the side toward the Sun while on the side opposite the Sun they are crowned with brighter borders like shining ridges. And we have an almost entirely similar sight on Earth, around sunrise, when the valleys are not yet bathed in light but the surrounding mountains facing the Sun are already seen shining with light. And just as the shadows of the earthly valleys are diminished as the Sun climbs higher, so those lunar spots lose their darkness as the luminous part grows.

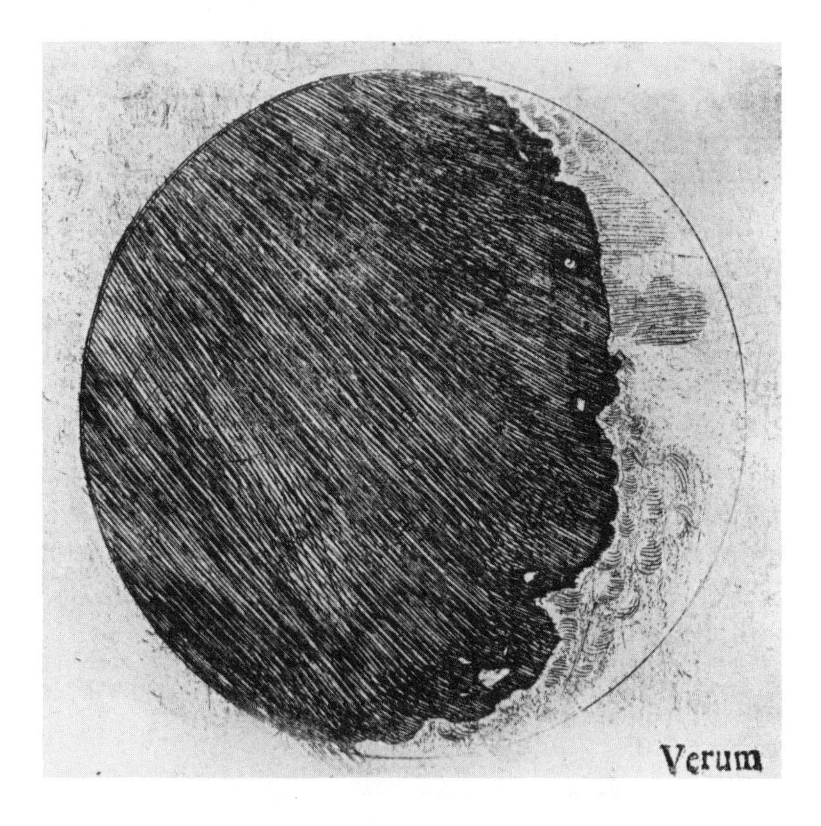

Not only are the boundaries between light and dark on the Moon perceived to be uneven and sinuous, but, what causes even greater wonder, is that very many bright points appear within the dark part of the Moon, entirely separated and removed from the illuminated region and located no small distance from it. Gradually, after a small period of time, these are increased in size and brightness. Indeed, after 2 or 3 hours they are joined with the rest of the bright part, which has now become larger. In the meantime, more and more bright points light up, as if they are sprouting, in the dark part, grow, and are connected at length with that bright surface as it extends farther in this direction. An example of this is shown in the same figure. Now, on Earth, before sunrise, aren't the peaks of the highest mountains illuminated by the Sun's rays while shadows still cover the plain? Doesn't light grow, after a little while, until the middle and larger parts of the same mountains are illuminated, and finally, when the Sun has risen, aren't the illuminations of plains and hills joined together? These differences between prominences and depressions in the Moon, however, seem to exceed the terrestrial roughness greatly, as we shall demonstrate below. Meanwhile, I would by no means be silent about something deserving notice, observed by me while the Moon was rushing toward first quadrature,[38] the appearance of which is also shown in the above figure. For toward the lower horn[39] a vast dark gulf projected into the bright part. As I observed this for a long time, I saw it very dark. Finally, after about 2 hours, a bit below the middle of this cavity a certain bright peak began to rise and, gradually growing, it assumed a triangular shape, still entirely removed

38. The Moon or a planet is at quadrature when its angular separation from the Sun is 90 degrees. The first quadrature of the Moon after new moon is called "first quarter."

39. On modern moonmaps, until recently, this would be the upper horn. While Galileo's telescope showed an erect (right-side up) image, modern instruments show an inverted (upside-down) image and for this reason modern Moon maps are drawn upside down. Note, however, that since spacecraft started sending back erect planetary images, more and more moonmaps are not inverted.

and separated from the bright face. Presently three other small points began to shine around it until, as the Moon was about to set, this enlarged triangular shape, now made larger, joined together with the rest of the bright part, and like a huge promontory, surrounded by the three bright peaks already mentioned, it broke out into the dark gulf. Also, in the tips of both the upper and lower horns, some bright points emerged, entirely separated from the rest of the light, as shown in the same figure. And there was a great abundance of dark spots in both horns, especially in the lower one. Of these, those closer to the boundary between light and dark appeared larger and darker while those farther away appeared less dark and more diluted. But as we have mentioned above, the dark part of the spot always faces the direction of the Sun and the brighter border surrounds the dark spot on the side turned away from the Sun and facing the dark part of the Moon. This lunar surface, which is decorated with spots like the dark blue eyes in the tail of a peacock, is rendered similar to those small glass vessels which, plunged into cold water while still warm, crack and acquire a wavy surface, after which they are commonly called ice-glasses. The large [and ancient] spots of the Moon, however, when broken up in a similar manner, are not seen to be filled with depressions and prominences, but rather to be even and uniform, for they are only here and there sprinkled with some brighter little places. Thus, if anyone wanted to resuscitate the old opinion of the Pythagoreans that the Moon is, as it were, another Earth, its brighter part would represent the land surface while its darker part would more appropriately represent the water surface.[40] Indeed, for me there has never been any doubt that when the terrestrial globe, bathed in sunlight, is observed from a distance, the land surface will present itself brighter to the view and the water surface darker. Moreover, in the Moon the large spots are seen to be lower than the brighter areas, for in her waxing as well as waning, on the border between light and dark, there is

40. For Kepler's discussion of this aspect, see p. 95, below.

always a prominence here or there around these large spots, next to the brighter part, as we have taken care to show in the figures; and the edges of the said spots are not only lower, but more uniform and not broken by creases or roughnesses. Indeed, the brighter part stands out very much near the ancient spots, so that both before the first and near the second quadrature some huge projections arise around a certain spot in the upper, northern part of the Moon, both above and below it, as the adjoining figures show.

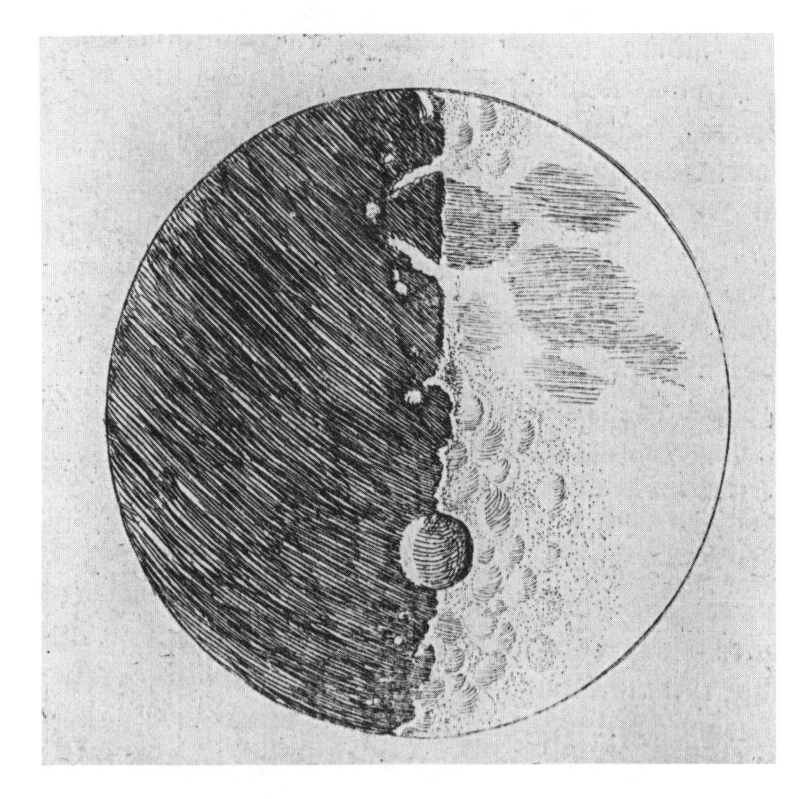

Before the second quadrature this same spot is seen walled around by some darker edges which, like a ridge of very high mountains turned away from the Sun, appear darker; and where they face the Sun they are brighter. The opposite of this occurs in valleys whose part away from the Sun appears brighter, while the part situated toward the Sun is dark and shady. Then, when the bright surface has decreased in size, as soon as almost this entire spot is covered in darkness, brighter ridges of mountains rise loftily out of the darkness. The following figures clearly demonstrate this double appearance.

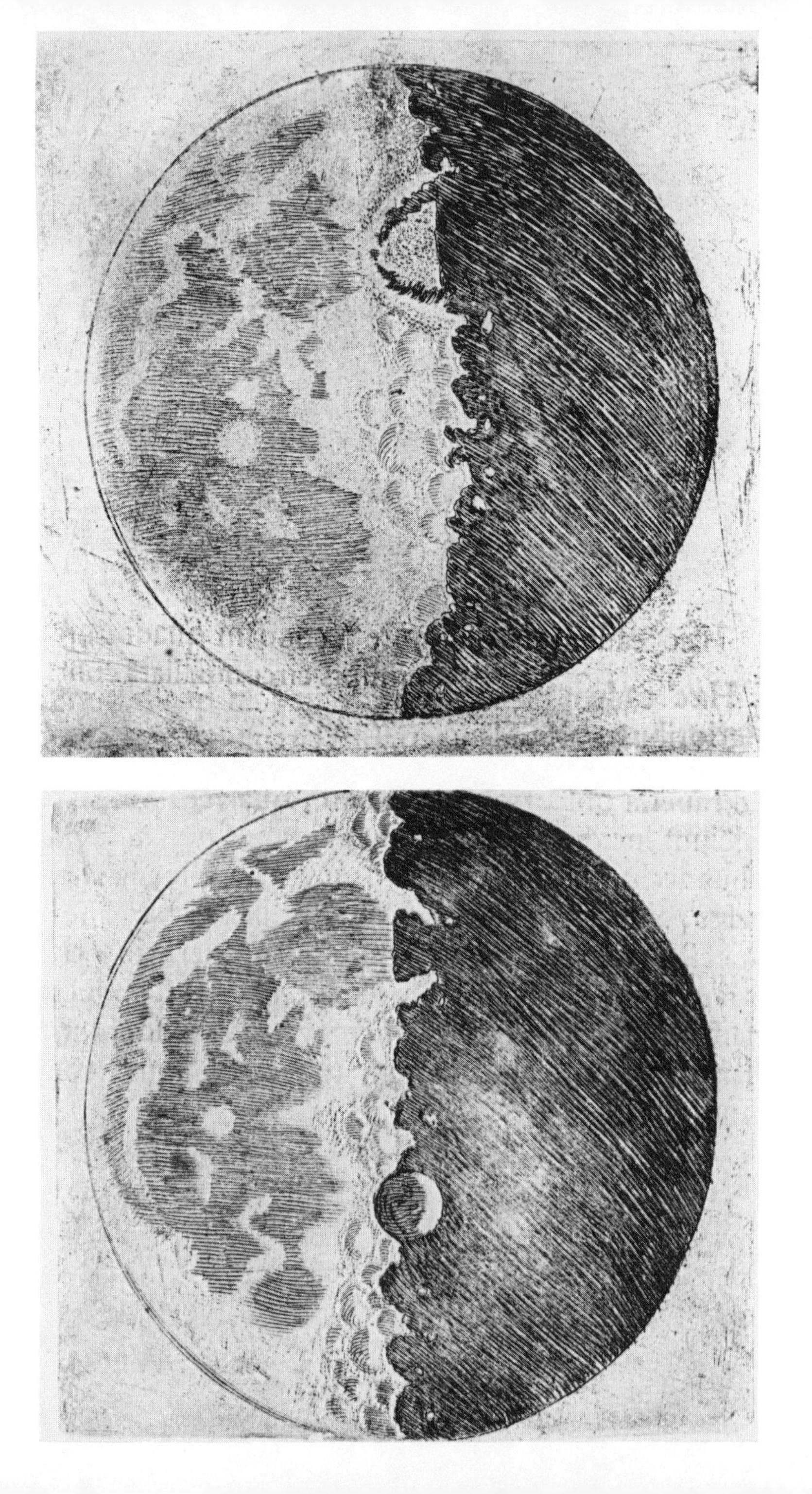

There is another thing that I noticed not without some admiration and that I may not omit. The area around the middle of the Moon is occupied by a certain cavity larger than all others and of a perfectly round figure.[41] I observed this near both quadratures, and I have portrayed it as far as possible in the second figure above. It offers the same aspect to shadow and illumination as a region similar to Bohemia would offer on Earth, if it were enclosed on all sides by very high mountains, placed around the periphery in a perfect circle. For on the Moon it is surrounded by such lofty ranges that its side bordering on the dark part of the Moon is observed bathed in sunlight before the dividing line between light and shadow reaches the middle of the diameter of that circle. But in the manner of the other spots, its shaded part faces the Sun while its bright part is situated toward the dark part of the Moon, which, I advise for the third time, is to be esteemed as a very strong argument for the roughnesses and unevennesses scattered over the entire brighter region of the Moon. Its darker spots are always those that border on the boundary between light and dark, while the farther ones appear both smaller and less dark, so that finally, when the Moon is at opposition and full, the darkness of the depressions differs from the brightness of the prominences by a modest and quite small degree.

These things we have reviewed are observed in the brighter regions of the Moon. In the large spots, however, such a difference between depressions and prominences is not seen to be the same, as we are driven to conclude by necessity in the brighter part on account of the change of shapes caused by the changing illumination of the Sun's rays as it regards the Moon from many different positions. In the large spots there are some darkish areas, as we have shown in the figures, but yet those always have the same appear-

41. It was not Galileo's purpose to make an accurate map of the Moon, but rather to illustrate its Earth-like nature. It is therefore often difficult to identify features on his drawings. In the case of this obviously exaggerated "cavity," the best guess is that it represents the crater Albategnius. For a discussion of this problem, see the articles cited in note 23, p. 9, above.

ance, and their darkness is not increased or abated. Rather, they appear, with a very slight difference, now a little darker, now a little lighter, as the Sun's rays fall on them more or less obliquely. Moreover, they join with nearby parts of the spots in a gentle bond, their boundaries mingling and running together. Things happen differently, however, in the spots occupying the brighter part of the Moon, for like sheer cliffs sprinkled with rough and jagged rocks, these are divided by a line which sharply separates shadow from light. Moreover, in those larger spots certain other brighter areas—indeed, some very bright ones—are seen. But the appearance of these and the darker ones is always the same, with no change in shape, light, or shadow. It is thus known for certain and beyond doubt that they appear this way because of a real dissimilarity of parts and not merely because of inequalities in the shapes of their parts and shadows moving diversely because of the varying illumination by the Sun. This does happen beautifully in the other, smaller, spots occupying the brighter part of the Moon; day by day these are altered, increased, diminished, and destroyed, since they only derive from the shadows of rising prominences.

But I sense that many people are affected by great doubt in this matter and are so occupied by the grave difficulty that they are driven to call into doubt the conclusion already explained and confirmed by so many appearances. For if that part of the Moon's surface which more brilliantly reflects the solar rays is filled with innumerable contortions, that is, elevations and depressions, why is it that in the waxing Moon the limb facing west, in the waning Moon the eastern limb, and in the full Moon the entire periphery are seen not uneven, rough, and sinuous, but exactly round and circular, and not jagged with prominences and depressions? And especially because the entire edge consists of the brighter lunar substance which, we have said, is entirely bumpy and covered with depressions, for none of the large [ancient] spots reach the very edge, but all are seen to be clustered far from the periphery. Since these appearances present an opportunity for such serious doubt, I shall put forward a double cause for them and therefore a double explanation of the doubt. First, if the prominences and depressions in the lunar body were spread only along the single circular periphery outlining the hemisphere seen by us, then the Moon could indeed, nay it would have to, show itself to us in the shape of, as it were, a toothed wheel, that is, bumpy and bounded by a sinuous outline. If, however, there were not just one chain of prominences distributed only along a single circumference, but rather very many rows of mountains with their clefts and sinuosities were arranged about the outer circuit of the Moon—and these not only in the visible hemisphere but also in the one turned away from us (yet near the boundary between the hemispheres)—then the eye, observing from afar, could by no means perceive the distinction between the prominences and depressions. For the interruptions in the mountains arranged in the same circle or the same chain are hidden by the interposition of row upon row of other prominences, and especially if the eye of the observer is located on the same line with the peaks of those prominences. Thus on Earth the ridges of many mountains close together appear to be arranged in a flat surface if the observer is far away and situated at the same altitude. So [also] in a billowy sea the high tips of the waves appear stretched out in the same plane, even though between the waves there are very many troughs and gulfs so deep that not only the keels but also the upper decks, masts, and sails of tall ships are hidden. Since, therefore, in the Moon itself and around its perimeter there is a complex arrangement of prominences and depressions, and the eye, observing from afar, is located in about the same plane as their peaks, it should be surprising to no one that, with the visual rays skimming them, they show themselves in an even and not at all wavy line.[42] To this reason another can be added, namely, that, just as around the Earth, there is around the lunar body a certain orb of denser substance than the rest of the ether, able to receive and

42. Although Galileo's argument is cogent, successive mountain ranges do not make the Moon's outline perfectly smooth. With modern instruments the remaining unevennesses can easily be observed.

reflect a ray of the Sun, although not endowed with so much opacity that it can inhibit the passage of vision (especially when it is not illuminated). That orb, illuminated by the solar rays, renders and represents the lunar body in the figure of a larger sphere, and, if it were thicker, it could limit our sight so as not to reach the actual body of the Moon. And it is indeed thicker around the periphery of the Moon; not absolutely thicker I say, but thicker as presented to our [visual] rays that intersect it obliquely. Therefore it can inhibit our vision and, especially when it is luminous, hide the periphery of the Moon exposed to the Sun. This is seen clearly in the adjoining figure, in which the lunar body *ABC* is surrounded by the vaporous

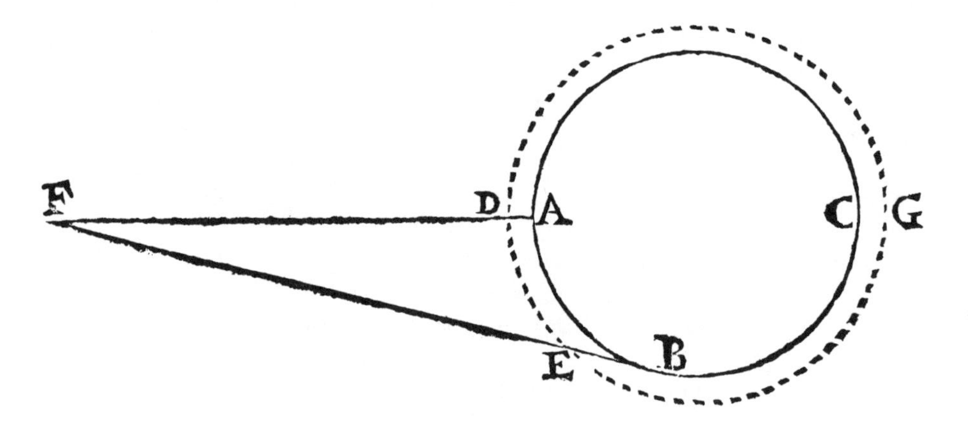

orb *DEG*. The eye at *F* reaches the middle parts of the Moon, as at *A*, through the shallower vapors *DA*, and toward its extreme parts an abundance of deeper vapors, *EB*, blocks our sight from its boundary. An indication of this is that the part of the Moon bathed in light appears greater in circumference than the remaining dark orb. And someone will perhaps find this cause reasonable to explain why the larger spots of the Moon are nowhere seen to extend to the outer edge, although it is to be expected that some of them would also be found near it. It seems plausible, then, that they are inconspicuous because they are hidden under thicker and brighter vapors.[43]

From the appearances already explained, I think it is sufficiently clear that the brighter surface of the Moon is sprinkled all over with prominences and depressions. It remains for us to speak of their magnitudes, demonstrating that the terrestrial roughnesses are far smaller than the lunar ones. I say smaller, speaking absolutely, not merely in proportion to the sizes of their globes. This is clearly shown in the following manner.

As has often been observed by me, with the Moon in various aspects to the Sun, some peaks within the dark part of the Moon appear drenched in light, although very far from the boundary line of the light. Comparing their distance from that boundary line to the entire lunar diameter, I found that this interval sometimes exceeds the twentieth part of the diameter. Assuming this, imagine the lunar globe, whose great circle is *CAF,* whose center is *E,* and whose diameter is *CF,* which is to the Earth's diameter as 2 to 7. And since according to the most exact observations the terrestrial diameter contains 7000 Italian miles,[44] *CF* will be 2000 miles, *CE* 1000, and the twentieth part of the whole of *CF* will be 100 miles. Now let *CF* be the diameter of the great circle dividing the luminous from the dark part of the Moon (because of the very great distance of the Sun from the Moon this circle does not differ sensibly from a great circle), and let *A* be distant from point *C* one-twentieth part of it. Draw the semidiameter *EA,* which, when extended, intersects the tangent *GCD* (which represents a ray of light) at *D.* The arc *CA* or the straight line *CD* will therefore be 100 parts of the 1000 represented by *CE,* and the sum of the squares of *CD* and *CE* is 1,010,000, which is equal to the square of *ED.* The whole

43. This argument was later abandoned by Galileo: it is not to be found in his *Dialogue concerning the Two Chief World Systems* of 1632, in which he treats the appearance of the Moon in great detail.

44. Galileo was using convenient numbers and fractions here. The diameters of the Earth and Moon had been known with surprising accuracy since Antiquity. See A. Van Helden, *Measuring the Universe,* 4–27.

of *ED* will therefore be more than 1004,[45] and *AD* more than 4 parts of the 1000 represented by *CE*. Therefore the height *AD* on the Moon, which represents some peak reaching all the way up to the Sun's rays *GCD* and removed from the boundary line *C* by the distance *CD*, is higher than 4 Italian miles.[46] But on Earth no mountains exist that reach even to a perpendicular height of 1 mile.[47]

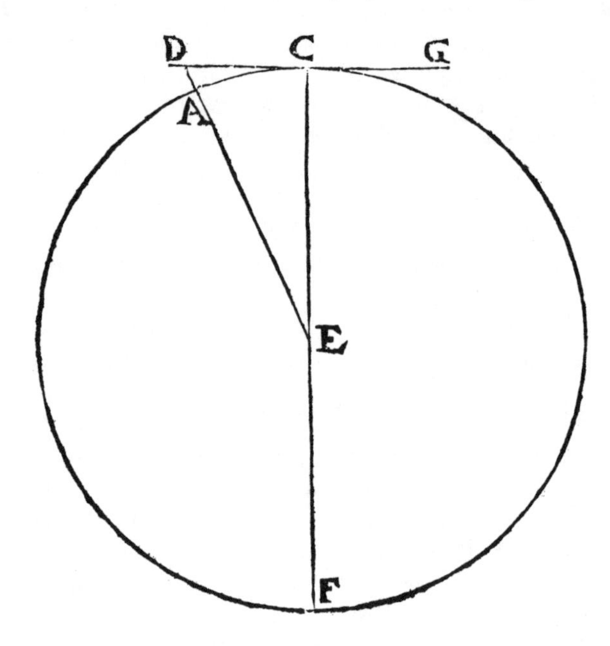

45. The square root of 1,010,000 is almost exactly 1005, which would make Galileo's argument even stronger.

46. There was no Italian mile strictly speaking. The miles of Florence, Venice, and Rome were all within 10% of the modern English mile.

47. Efforts to determine the heights of mountains by geometrical means began in Greek Antiquity. These first estimates were generally of the order of a mile. Later speculations varied widely, and it appears that Galileo relied on earlier sources. See Florian Cajori, "History of Determinations of the Heights of Mountains," *Isis* 12 (1929): 482–514; and C. W. Adams, "A Note on Galileo's Determination of the Height of Lunar Mountains," *Isis* 17 (1932): 427–29.

It is evidence, therefore, that the lunar prominences are loftier than the terrestrial ones.

In this place I wish to explain the cause of another lunar phenomenon worthy of notice. This phenomenon was observed by us not recently but rather many years ago, shown to some close friends and pupils, explained, and given a causal demonstration. But since the observation of it is made easier and more noticeable with the aid of the glasses, I thought it not unsuitable to be repeated here, and especially so that the relationship and similarity between the Moon and Earth may appear more clearly.

When, both before and after conjunction,[48] the Moon is found not far from the Sun, she offers to our sight not only that part of her globe that is adorned with shining horns, but also a certain thin, faint periphery that is seen to outline the circle of the dark part (that is, the part turned away from the Sun) and to separate it from the darker field of the ether itself. But if we examine the matter more closely, we will see not only the extreme edge of the dark part shining with a faint brightness, but the entire face of the Moon—that part, namely, that does not yet feel the brightness of the Sun—made white by some not inconsiderable light.[49] At first glance, however, only a slender shining circumference appears on account of the darker parts of the sky bordering it, while, on the contrary, the rest of the surface appears darker because the nearness of the shining horns makes our sight dark. But if one chooses a place for oneself so that those bright horns are concealed by a roof or a chimney or another obstacle between one's sight and the Moon (but positioned far away from the eye), the remaining part of the lunar globe is left exposed to one's view, and then one will discover that this region of the Moon, although deprived of sunlight, also shines with a considerable light, and especially when the chill of the night has already increased through the absence of the Sun. For in a darker field the same light appears brighter. It is moreover ascertained that this secondary

48. With the Sun.

49. This phenomenon was called *lumen cinereum* or "ashen light" of the Moon. It is now referred to as Earth-shine. See p. 22, above.

brightness (as I call it) of the Moon is greater the less distant the Moon is from the Sun, for as she becomes more distant from him it is decreased more and more so that after the first quadrature and before the second it is found weak and very doubtful, even though it is seen in a darker sky, while at the sextile[50] and smaller elongations it shines in a wonderful way although in the twilight. Indeed, it shines so much that with the aid of a precise glass the large spots can be distinguished in her. This marvelous brightness has caused no small astonishment to those applying themselves to philosophy, and some have put forward one reason and some another as the cause to be assigned to it. Some have said that it is the intrinsic and natural brightness of the Moon herself;[51] others that it is imparted to it by Venus, or by all the stars;[52] and yet others have said that it is imparted by the Sun who penetrates the Moon's vast mass with his rays.[53] But such inventions are refuted with little difficulty and demonstrated to be false. For if this kind of light were either the Moon's own or gathered from the stars, she would retain it and show it especially during eclipses when she is placed in a very dark sky. This is not borne out by experience, however, for the light that appears in the Moon during an eclipse is much weaker, somewhat reddish, and almost coppery,[54] while this light is brighter and whiter. The light that

50. That is, when the angular separation between Sun and Moon is 60 degrees.

51. For example, Erasmus Reinhold. See his edition of Peurbach's *Theoricae Novae Planetarum* (Wittenberg, 1553), ff. $164^{v}-165^{r}$. See also Kepler, *Gesammelte Werke,* 2:221–22.

52. This was maintained by Tycho Brahe, in his *Astronomiae Instauratae Progymnasmata* (1602), according to Kepler. See *Gesammelte Werke,* 2:223; and Rosen, *Kepler's Conversation,* 119–20.

53. Vitello, *Perspectiva,* iv, 77. See *Opticae Thesaurus,* ed. Friedrich Risner (Basel, 1572; reprint, New York: Johnson Reprint Corp., 1972), p. 151 of *Vitellionis Opticae.*

54. The reddish color of the Moon during a lunar eclipse is now attributed to refraction of sunlight by the Earth's atmosphere. Sunlight that grazes the Earth is refracted and illuminates the Moon slightly. But in the passage through the Earth's atmosphere the wavelengths at the blue end of the spectrum are scattered, so that only wavelengths near the red end of the spectrum pass through. This same absorption makes the Sun appear red at sunrise and sunset.

appears during an eclipse is, moreover, changeable and movable, for it wanders across the lunar face so that the part closer to the edge of the circle of the Earth's shadow is always seen brighter while the rest is darker. From this we understand with complete certainty that this light comes about because of the proximity of the solar rays falling upon some denser region which surrounds the Moon on all sides. Because of this contact a certain dawn light is spread over nearby areas of the Moon, just as on Earth twilight is spread in the morning and evening. We will treat this matter at greater length in a book on the system of the world.[55] To declare, on the other hand, that this light is imparted by Venus is so childish as to be unworthy of an answer. For who is so ignorant as not to know that near conjunction and within the sextile aspect it is entirely impossible for the part of the Moon turned away from the Sun to be seen from Venus? But it is equally inconceivable that this light is due to the Sun, who with his light penetrates and fills the solid body of the Moon. For it would never be diminished, since a hemisphere of the Moon is always illuminated by the Sun except at the moment of a lunar eclipse. Yet the light is diminished when the Moon hastens toward quadrature and is entirely dimmed when she has gone beyond quadrature. Since, therefore, this secondary light is not intrinsic and proper to the Moon, and is borrowed neither from any star nor from the Sun, and since in the vastness of the world no other body therefore remains except the Earth, I ask what are we to think? What are we to propose—that the lunar body or some other dark and gloomy body is bathed by light from the Earth? But what is so surprising about that? In an equal and grateful exchange the Earth pays back the Moon with light equal to that which she receives from the Moon almost all the time in the deepest darkness of the night. Let us demonstrate the matter more clearly. At conjunction, when she occupies a place between the Sun and the Earth, the Moon is flooded by solar rays on her upper hemisphere that is turned away from the Earth. But the lower hemisphere turned toward the Earth is covered in darkness, and therefore

55. See *Dialogue concerning the Two Chief World Systems*, 67–99.

it in no way illuminates the terrestrial surface. As the Moon gradually recedes from the Sun, some part of the inferior hemisphere turned toward us is soon illuminated and she turns somewhat white but thin horns toward us and lightly illuminates the Earth. The illumination of the Sun grows on the Moon as she approaches quadrature, and on Earth the reflection of her light increases. The brightness of the Moon is extended further, beyond a semicircle, and lights up our clear nights. Finally, the entire face of the Moon that regards the Earth is illuminated with a very bright light from the opposed Sun, and the Earth's surface shines far and wide, perfused by lunar splendor. Afterward, when the Moon is waning, she emits weaker rays toward us, and the Earth is weakly illuminated; and as the Moon hastens toward conjunction, dark night comes over the Earth. In this sequence, then, in alternate succession, the lunar light bestows upon us her monthly illuminations, now brighter, now weaker. But the favor is repaid by the Earth in like manner, for when the Moon is found near the Sun around conjunction, she faces the entire surface of the hemisphere of the Earth exposed to the Sun and illuminated by vigorous rays, and receives reflected light from her. And therefore, because of this reflection, the inferior hemisphere of the Moon, although destitute of solar light, appears of considerable brightness. When the Moon is removed from the Sun by a quadrant, she only sees the illuminated half of the terrestrial hemisphere, that is, the western one, for the other, the eastern half, is darkened by night. The Moon is therefore less brightly illuminated by the Earth, and her secondary light accordingly appears more feeble to us. For if you place the Moon at opposition to the Sun, she will face the hemisphere of the interposed Earth that is entirely dark and steeped in the shadow of night. If therefore, such an opposition were an eclipse, the Moon would receive absolutely no illumination, being deprived alike of solar and terrestrial radiation. In its various aspects to the Sun and Earth, the Moon receives more or less light by reflection from the Earth as she faces a larger or smaller part of the illuminated terrestrial hemisphere. For the relative positions of those two globes are always such that at those times when the Earth is most illuminated by the Moon the Moon is least illuminated by the Earth, and vice versa. Let these few things said here about this matter suffice. We will say more in our *System of the World,* where with very many arguments and experiments a very strong reflection of solar light from the Earth is demonstrated to those who claim that the Earth is to be excluded from the dance of the stars, especially because she is devoid of motion and light. For we will demonstrate that she is movable and surpasses the Moon in brightness, and that she is not the dump heap of the filth and dregs of the universe, and we will confirm this with innumerable arguments from nature.[56]

Up to this point we have discussed the observations made of the lunar body. We will now report briefly on what has been observed by us thus far concerning the fixed stars. And first, it is worthy of notice that when they are observed by means of the spyglass, stars, fixed as well as wandering, are seen not to be magnified in size in the same proportion in which other objects, and also the Moon herself, are increased.[57] In the stars,[58] the increase appears much smaller so that you may believe that a glass capable of multiplying other objects, for example, by a ratio of 100 hardly multiplies stars by a ratio of 4 or 5. The reason for this is that when the stars are observed with the naked eye, they do not show themselves according to their simple and, so to speak, naked size, but rather surrounded by a certain brightness and crowned by twinkling rays, especially as the night advances. Because of this they appear much larger than if they were stripped of these extraneous rays, for the visual angle is determined not by the primary body of the star but by the widely surrounding brilliance. You will perhaps understand this more clearly from this: that stars emerging in the first twilight at sunset, even if they are of the first magnitude, appear very small,

56. Ibid.

57. The first estimates of the apparent diameters of fixed stars and planets were made in Antiquity. These estimates, which were much too high, were faithfully followed by all successors of Ptolemy until the telescope showed them to be in error. See A. Van Helden, *Measuring the Universe.*

58. That is, fixed stars as well as planets.

and Venus herself, when she presents herself to our view in broad daylight,[59] is perceived so small that she hardly appears to equal a little star of the sixth magnitude. Things are different for other objects and the Moon herself, which, whether she is observed at midday or in the deepest darkness, appears always of the same size to us. Stars are therefore seen unshorn in the midst of darkness, but daylight can shear them of their hair—and not only daylight but also a thin little cloud that is interposed between the star and the eye of the observer. The same effect is also achieved by dark veils and colored glasses, by the opposition and interposition of which the surrounding brightness will desert the stars. The spyglass likewise does the same thing: for first it takes away the borrowed and accidental brightness from the stars and thereupon it enlarges their simple globes (if indeed their figures are globular), and therefore they appear increased by a much smaller ratio, for stars of the fifth or sixth magnitude seen through the spyglass are shown as of the first magnitude.[60]

The difference between the appearance of planets and fixed stars also seems worthy of notice. For the planets present entirely smooth and exactly circular globes that appear as little moons, entirely covered with light, while the fixed stars are not seen bounded by circular outlines but rather as pulsating all around with certain bright rays.[61] With the glass they appear in the same shape as when they are observed with natural vision, but so much larger that a little star of the fifth or sixth magnitude appears to equal the Dog

59. The Latin is *circa meridiem,* that is, around noon. Only on rare occasions, when Venus is at its greatest elongation from the Sun (about 45 degrees) and therefore at its brightest and when the seeing conditions are very good, can an observer with keen sight who knows exactly where to look see Venus with the naked eye around noon. Such occasions are so infrequent, however, that it is more likely Galileo simply meant an hour or so after sunrise or before sunset.

60. On Galileo's argument here, see Harold I. Brown, "Galileo on the Telescope and the Eye," *Journal for the History of Ideas* 46 (1985): 487–501.

61. Up to this point the only observable differences between fixed stars and planets had been in their motions and in the fact that the former twinkle while the latter do not.

Star,[62] which is the largest of all fixed stars. Indeed, with the glass you will detect below stars of the sixth magnitude such a crowd of others that escape natural sight that it is hardly believable. For you may see more than six further gradations of magnitude. The largest of these, which we may designate as of the seventh magnitude, or the first magnitude of the invisible ones, appear larger and brighter with the help of the glass than stars of the second magnitude seen with natural vision.[63] But in order that you may see one or two illustrations of the almost inconceivable crowd of them, and from their example form a judgment about the rest of them, I decided to reproduce two star groups. In the first I had decided to depict the entire constellation of Orion, but overwhelmed by the enormous multitude of stars and a lack of time, I put off this assault until another occasion.[64] For there are more than five hundred new stars around the old ones, spread over a space of 1 or 2 degrees. For this reason, to the three in Orion's belt and the six in his sword[65] that were observed long ago, I have added eighty

62. Sirius.

63. These figures present a problem. A difference of 5 magnitudes in brightness represents a factor of 100 in the amount of light gathered, or a tenfold increase in aperture. The light gathered by the eye is governed by the aperture of the pupil, and the dark-adapted pupil has a diameter of about $\frac{1}{3}$ inch. This would mean that Galileo's spyglass had an aperture of well over 3 inches, and we know this was not the case. The apertures of his instruments were stopped down to 1 inch or less. We can only conclude, therefore, that when Galileo made this estimate, his pupil was not yet adapted to the dark, and was therefore considerably smaller than $\frac{1}{3}$ inch.

64. Nothing in his papers suggests that Galileo ever made a map of the entire constellation.

65. Galileo does not show the nebula in the sword of Orion, which is a naked-eye object. It is registered as a star, without the qualification "cloudy," in the star catalogs of Ptolemy and Copernicus. For this reason it has been suggested that this nebula has changed during historical time: see Thomas G. Harrison, "The Orion Nebula: Where in History Is It?" *Quarterly Journal of the Royal Astronomical Society* 25 (1984): 65–79. For an assessment of this argument, see Owen Gingerich, "The Mysterious Nebulae, 1610–1924," *Journal of the Royal Astronomical Society of Canada* 81 (1987): 113–27. The nebula was first observed by Peiresc in 1611: see Pierre Humbert, *Un amateur: Peiresc, 1580–1637* (Paris: *Desclée, de Brouwer et cie.*, 1933),

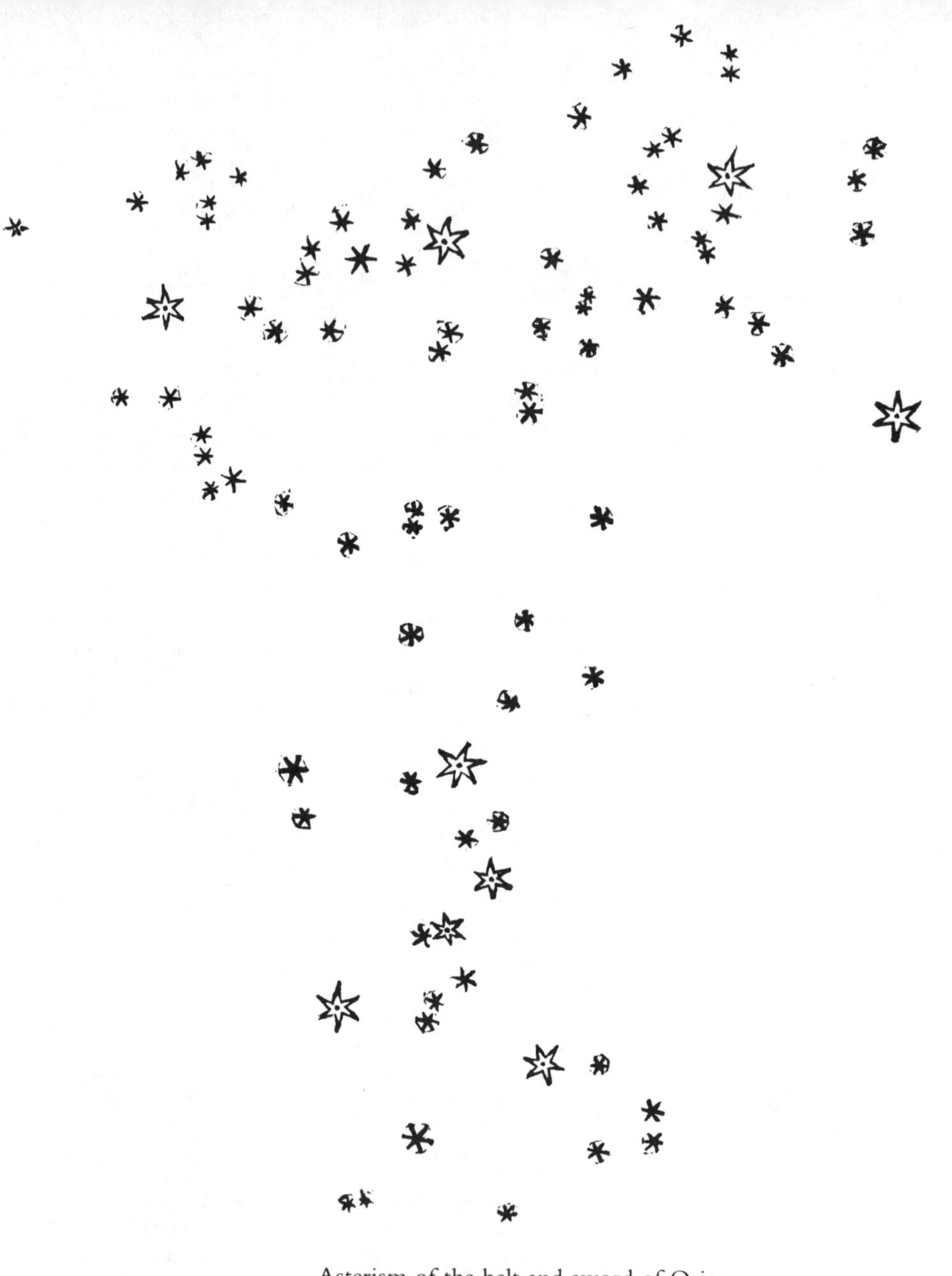

Asterism of the belt and sword of Orion

Constellation of the Pleiades

others seen recently, and I have retained their separations as accurately as possible. For the sake of distinction, we have depicted the known or ancient ones larger and outlined by double lines, and the other inconspicuous ones smaller and outlined by single lines. We have also preserved the distinction in size as much as possible. In the second example we have depicted the six stars of the Bull[66] called the Pleiades (I say six since the seventh almost never appears)[67]

---

p. 42, and Seymour L. Chapin, "The Astronomical Activities of Nicolas Claude Fabri de Peiresc," *Isis* 48 (1957): 19–20. Note that the section containing the descriptions of star formations was added at a very late stage, for the four pages containing them are added between pp. 16ᵛ and 17ʳ and are unnumbered. We may surmise that Galileo was silent on this nebula because he was convinced that it could be resolved into individual stars with more powerful instruments and in the meantime did not wish to vitiate his argument.

66. Taurus.

67. The Pleiades group is an open cluster consisting of several thousand stars, about 400 light-years from Earth. Six of its stars are brighter than the fifth magnitude, and a total of nine are brighter than the sixth magnitude. With the naked eye observers therefore see either six or nine (and sometimes even more), depending on their eyesight, but never seven.

contained within very narrow limits in the heavens. Near these lie more than forty other invisible stars, none of which is farther removed from the aforementioned six than scarcely half a degree. We have marked down only thirty-six of these, preserving their mutual distances, sizes, and the distinction between old and new ones, as in the case of Orion.

What was observed by us in the third place is the nature or matter of the Milky Way itself, which, with the aid of the spyglass, may be observed so well that all the disputes that for so many generations have vexed philosophers are destroyed by visible certainty, and we are liberated from wordy arguments.[68] For the Galaxy is nothing else than a congeries of innumerable stars distributed in clusters. To whatever region of it you direct your spyglass, an immense number of stars immediately offer themselves to view, of which very many appear rather large and very conspicuous but the multitude of small ones is truly unfathomable.

And since that milky luster, like whitish clouds, is seen not only in the Milky Way, but dispersed through the ether, many similarly colored patches shine weakly; if you direct a glass to any of them, you will meet with a dense crowd of stars. Moreover—and what is even more remarkable—the stars that have been called "nebulous" by every single astronomer up to this day are swarms of small stars placed exceedingly closely together.[69] While each individual one escapes our sight because of its smallness or its very great distance from us, from the commingling of their rays arises that brightness ascribed up to now to a denser part of the heavens capable of reflecting the rays of the stars or Sun.[70] We have observed some of

68. For a review of pre-Galilean notions concerning the Milky Way, see Stanley L. Jaki, *The Milky Way: An Elusive Road for Science* (New York: Science History Publications; Newton Abbot: David & Charles, 1973), 1–101.

69. The six nebulous stars listed in Ptolemy's star catalog, and the five listed by Copernicus, can, in fact, all be resolved into stars. As it turned out, there is nebular matter in the universe. The question was not settled, however, until the advent of spectroscopy in the second half of the nineteenth century.

70. This notion was first put forward by Albertus Magnus in the thirteenth century; see Jaki, *The Milky Way*, 41. It was the explanation given by Christopher

these, and we wanted to reproduce the asterisms of two of them.

In the first there is the nebula called Orion's Head, in which we have counted twenty-one stars.[71]

The second figure contains the nebula called Praesepe, which is not a single star but a mass of more than forty little stars. In addition to the ass-colts we have marked down thirty-six stars, arranged as follows:[72]

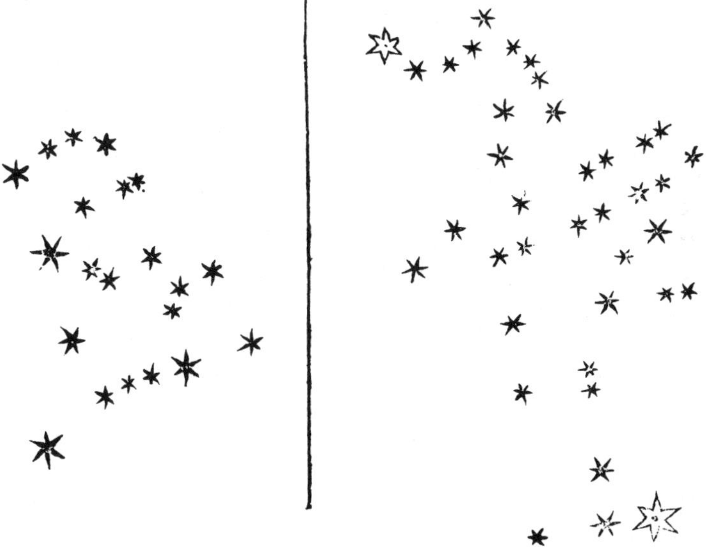

Nebula of Orion Nebula of Praesepe

Clavius (1537–1612) in his influential *Commentary on the Sphere of Sacrobosco* (1570), which went through numerous editions during Galileo's lifetime. See *In Sphaeram Ioannis de Sacro Bosco Commentarius* (Rome, 1570), 376–77.

71. This is the area near λ, ϕ^1, and ϕ^2 Orionis. Galileo no doubt chose this area because it is listed in Ptolemy's star catalog as nebulous. See *Ptolemy's Almagest,* ed. G. J. Toomer (London: Duckworth, 1984), 382.

72. The two large stars depicted here are γ and δ Cancri, called the *Aselli,* or ass-colts, in Antiquity. The nebulous area between them, NGC 2632 = M44, is Praesepe, the Manger or Beehive. It is listed by Ptolemy as nebulous. See *Ptolemy's Almagest,* 366.

We have briefly explained our observations thus far about the Moon, the fixed stars, and the Milky Way. It remains for us to reveal and make known what appears to be most important in the present matter: four planets never seen from the beginning of the world right up to our day, the occasion of their discovery and observation, their positions, and the observations made over the past 2 months[73] concerning their behavior and changes. And I call on all astronomers to devote themselves to investigating and determining their periods. Because of the shortness of time, it has not been possible for us to achieve this so far.[74] We advise them again, however, that they will need a very accurate glass like the one we have described at the beginning of this account, lest they undertake such an investigation in vain.[75]

Accordingly, on the seventh day of January of the present year 1610,[76] at the first hour of the night, when I inspected the celestial constellations through a spyglass, Jupiter presented himself. And since I had prepared for myself a superlative instrument, I saw (which earlier had not happened because of the weakness of the other instruments)[77] that three little stars were positioned near him—small but yet very bright. Although I believed them to be among the number of fixed stars, they nevertheless intrigued me because they appeared to be arranged exactly along a straight line and parallel to the ecliptic, and to be brighter than others of equal size. And their disposition among themselves and with respect to Jupiter was as follows:[78]

73. 7 January to 2 March 1610.

74. In 1612 Galileo published periods for all four satellites. They were virtually the same as the modern values. See *Discourse on Bodies in Water,* tr. Thomas Salusbury, ed. Stillman Drake (Urbana: University of Illinois Press, 1960), 1.

75. Especially in the case of the satellites of Jupiter, it was necessary to have a telescope that magnified fifteen times or more and was especially adapted for celestial use.

76. All dates used by Galileo are Gregorian.

77. See pp. 13–14, above.

78. Satellites I and II were very close together, just to the east of Jupiter. Galileo saw them as one. See Jean Meeus, "Galileo's First Records of Jupiter's Satellites," *Sky and Telescope* 24 (1962): 137–39.

East * * O * West

That is, two stars were near him on the east and one on the west; the more eastern one and the western one appeared a bit larger than the remaining one. I was not in the least concerned with their distances from Jupiter, for, as we said above, at first I believed them to be fixed stars. But when, on the eighth, I returned to the same observation, guided by I know not what fate,[79] I found a very different arrangement. For all three little stars were to the west of Jupiter and closer to each other than the previous night, and separated by equal intervals, as shown in the adjoining sketch.[80] Even though at this point I had by no means turned my thought to the mutual motions of these stars, yet I was aroused by the question

East O * * * West

of how Jupiter could be to the east of all the said fixed stars when the day before he had been to the west of two of them. I was afraid, therefore, that perhaps, contrary to the astronomical computations, his motion was direct and that, by his proper motion, he had bypassed those stars.[81] For this reason I waited eagerly for the next night. But I was disappointed in my hope, for the sky was everywhere covered with clouds.

Then, on the tenth, the stars appeared in this position with regard to Jupiter. Only two stars were near him, both to the east. The

East * * O West

79. See pp. 12–13, above.

80. On this night, satellite IV was at its farthest distance from Jupiter to the east, and it escaped Galileo because of the smallness of the field of view of his spyglass. See Meeus, "Galileo's First Records."

81. See pp. 15–16, above.

third, as I thought, was hidden behind Jupiter.[82] As before, they were in the same straight line with Jupiter and exactly aligned along the zodiac. When I saw this, and since I knew that such changes could in no way be assigned to Jupiter, and since I knew, moreover, that the observed stars were always the same ones (for no others, either preceding or following Jupiter, were present along the zodiac for a great distance), now, moving from doubt to astonishment, I found that the observed change was not in Jupiter but in the said stars. And therefore I decided that henceforth they should be observed more accurately and diligently.

And so, on the eleventh, I saw the following arrangement:

East * * O West

There were only two stars on the east,[83] of which the middle one was three times as far from Jupiter than from the more eastern one, and the more eastern one was about twice as large as the other, although the previous night they had appeared about equal. I therefore arrived at the conclusion, entirely beyond doubt, that in the heavens there are three stars wandering around Jupiter like Venus and Mercury around the Sun. This was at length seen clear as day in many subsequent observations, and also that there are not only three, but four wandering stars making their revolutions about Jupiter. The following is an account of the changes in their positions, accurately determined from then on. I also measured the distances between them with the glass, by the procedure explained above.[84] I have added the times of the observations, especially when more than one were made on the same night, for the revolutions of these

82. On this night, Satellite I was so close to Jupiter on the west that it was lost in the planet's glare. Satellites II and III were very close to each other and Galileo saw them as one, just to the east of the planet. See Meeus, "Galileo's First Records."

83. Satellites I and II had just ended their transits in front of the planet and were still too close to be discerned by Galileo. See ibid.

84. See pp. 38–39.

planets are so swift that the hourly differences can often be perceived as well.

Thus, on the twelfth, at the first hour of the following night, I saw the stars arranged in this manner. The more eastern star was

East  West

larger than the western one, but both were very conspicuous and bright.[85] Both were two minutes[86] distant from Jupiter. In the third hour a third little star, not at all seen earlier, also began to appear. This almost touched Jupiter on the eastern side and was very small. All were in the same straight line and aligned along the ecliptic.

On the thirteenth, for the first time four little stars were seen by me in this formation with respect to Jupiter.[87] Three were on

East * O * * * West

the west and one on the east. They formed a very nearly straight line, but the middle star of the western ones was displaced a little to the north from the straight line. The more eastern one was 2 minutes distant from Jupiter; the intervals between the remaining ones and Jupiter were only 1 minute. All these stars displayed the

85. Note that Galileo initially saw only two of the satellites, III on the east and II on the west. Satellites I and IV were both on the east and rather close to Jupiter. Apparently Galileo could not see either one until satellite I moved farther away from the planet. See Meeus, "Galileo's First Records."

86. Galileo took Jupiter's angular diameter to be about 1 arcminute, and he used this measure to estimate the distances of the satellites. In his drawings and in *Sidereus Nuncius,* however, he showed the planet's disk as being about twice as large while keeping the distances of the satellites the same. The drawings are thus out of proportion. See Stillman Drake, *Telescopes, Tides, and Tactics* (Chicago: University of Chicago Press, 1983), 214–19.

87. It was thus on this day that Galileo recognized that there were four moons. During the previous observations he had been prevented by various circumstances from seeing all four moons at once.

same size, and although small they were nevertheless very brilliant and much brighter than fixed stars of the same size.

On the fourteenth, the weather was cloudy.

On the fifteenth, in the third hour of the night, the four stars were positioned with respect to Jupiter as shown in the next figure.

East ○ * * * * West

They were all to the west and arranged very nearly in a straight line, except that the third one from Jupiter was raised a little bit to the north. The closest one to Jupiter was the smallest of all, and the rest consequently appeared larger. The intervals between Jupiter and the next three stars were all equal and of 2 minutes; and the most eastern one was 4 minutes from the closest one to it. They were very brilliant and did not twinkle, as indeed was always the case, both before and afterward. But in the seventh hour only three stars were present in this arrangement with Jupiter. They were

East

indeed precisely[88] in the same straight line. The closest one to Jupiter was very small and removed from him by 3 minutes; the second was 1 minute distant from this one; and the third from the second 4 minutes and 30 seconds. After another hour, however, the two little stars in the middle were still closer to each other, for they were removed from each other by barely 30 seconds.

On the sixteenth, in the first hour of the night, we saw three stars arranged in this order. Two flanked Jupiter, 40 seconds re-

East *○* * West

moved from him on either side, and the third was 8 minutes from Jupiter in the west. The one closer to Jupiter appeared not larger but brighter than the farther one.

88. I have translated the Latin *ad unguem* as *precisely* throughout this section.

On the seventeenth, 30 minutes after sunset, the configuration was thus. There was only one star on the east, 3 minutes from

East  * West

Jupiter. Likewise, one was 11 minutes from Jupiter to the west. The eastern one appeared twice as large as the western one. There were no more than these two. But after 4 hours, that is, around the fifth hour of the night, on the eastern side a third began to emerge, which, I suspect, had earlier been united with the first one. The formation was thus. The middle star, extremely close to the

East * * O * West

eastern one, was only 20 seconds from it, and it was displaced a little bit to the south of the line drawn through the outermost stars and Jupiter.

On the eighteenth, 20 minutes after sunset, the appearance was

East * O * West

thus. The eastern star was larger than the western one and 8 minutes distant from Jupiter, while the western one was 10 minutes from Jupiter.

On the nineteenth, at the second hour of the night, the formation was like this. There were three stars exactly on a straight line

East * O * * West

through Jupiter, one to the east, 6 minutes distant; between Jupiter and the first western one was an interval of 5 minutes, while this star was 4 minutes from the more western one. At this time I was uncertain whether between the eastern star and Jupiter there was a little star, very close to Jupiter, so that it almost touched him. And at the fifth hour, I clearly saw this little star now occupying a place precisely in the middle between Jupiter and the eastern star, so that the formation was as follows:

East West

Further, the newly perceived star was very small; yet by the sixth hour it was almost equal in magnitude to the others.

On the twentieth, at 1 hour, 15 minutes, a similar configuration appeared. There were three little stars so small that they could

East West

hardly be perceived. They were not more than 1 minute from Jupiter and each other. I was uncertain whether on the west there were two or three little stars. Around the sixth hour they were arranged in this manner:

East West

The eastern one was twice as far from Jupiter as before, that is, 2 minutes; the middle one to the west was 40 seconds from Jupiter but 20 seconds from the western one. At length, in the seventh hour, three little stars were seen to the west; the nearest was 20 seconds from him; between this one and the westernmost one there

East West

was an interval of 40 seconds; and between these another was seen, displaced a little to the south and not more than 10 seconds from the westernmost one.

On the twenty-first, at 30 minutes, there were three little stars to the east, equally spaced from each other and Jupiter.

East

The intervals were estimated to be 50 seconds. There was also a star to the west, 4 minutes from Jupiter. The closest one to Jupiter to the east was the smallest of all. The rest were somewhat larger and equal to each other.

On the twenty-second, at the second hour, the configuration was similar. The distance from the eastern one to Jupiter was 5 minutes; the distance from Jupiter to the westernmost one was 7

East * O ** * West

minutes; the two western stars in the middle were 40 seconds from each other while the nearer one was 1 minute from Jupiter. The little stars in the middle were smaller than the outermost ones, but they were on the same straight line extended along the length of the zodiac except that of the three western ones the middle one was displaced a bit to the south. But at the sixth hour of the night they appeared in this arrangement. The eastern one was very small and, as before, 5 minutes distant from Jupiter; the three western

East * O * * * West

ones were separated equally from Jupiter and each other, and the spaces were nearly 1 minute, 20 seconds each; and the star closer to Jupiter than the other two appeared smaller; and they all appeared to lie exactly on the same straight line.

On the twenty-third, 40 minutes after sunset, the configuration of stars was about like this:

East * * O * West

There were three stars in a straight line with Jupiter along the length of the zodiac, as they have always been; two were to the east and one to the west. The easternmost one was 7 minutes from the next one, this one 2 minutes, 40 seconds from Jupiter, and Jupiter 3 minutes, 20 seconds from the western one; and they were all about equal in magnitude. But at the fifth hour the two stars which earlier were closest to Jupiter were no longer visible, hiding behind Jupiter in my opinion; and the appearance was as follows:

East * O West

On the twenty-fourth, three stars appeared, all to the east, and nearly in the same straight line with Jupiter, for the middle one

East * ** O West

deviated slightly to the south. The star closest to Jupiter was 2 minutes from him, the next one 30 seconds from this one, and the easternmost one 9 minutes from that one; and all were very bright. But at the sixth hour only two stars presented themselves in this arrangement,

East [*] * O West

that is, precisely on a straight line with Jupiter. The nearer one was removed from Jupiter by 3 minutes while the other one was 8 minutes from this one. If I am not mistaken, the two middle little stars observed earlier had united into one.

On the twenty-fifth, at 1 hour, 40 minutes, the formation was thus:

East * * O West

There were only two stars to the east, and those fairly large. The easternmost was 5 minutes from the middle one, and the middle one 6 minutes from Jupiter.

On the twenty-sixth, at 0 hours, 40 minutes, the formation of stars was like this. For three stars were observed, of which two

were to the east and one to the west. This last one was 5 minutes from him, while the middle one in the east was 5 minutes, 20 seconds from him. The easternmost was 6 minutes from the middle one. They were on the same straight line and of the same magnitude. Then at the fifth hour the arrangement was nearly the same, differing only in this, that near Jupiter a fourth star had emerged

East * * *O * West

on the east, smaller than the rest, at that time 30 seconds removed from Jupiter but elevated a little to the north above the straight line, as shown in the adjoining figure.

On the twenty-seventh, at 1 hour after sunset, only a single star was perceived, and that one to the east, in this arrangement:

East * O West

It was very small and 7 minutes removed from Jupiter.

On the twenty-eighth and twenty-ninth, nothing could be observed because of interposed clouds.

On the thirtieth, at the first hour of the night, the stars were observed arranged in this order. One was to the east, 2 minutes,

East * O * * West

30 seconds from Jupiter, and two were to the west, of which the one closest to Jupiter was 3 minutes from him and the other 1 minute from this one. The outermost stars and Jupiter were arranged in a straight line, and the middle star was elevated a little to the north. The westernmost star was smaller than the others.

On the last day [of January], at the second hour, two stars appeared to the east and one to the west. The middle of the eastern

East ** O * West

ones was 2 minutes, 20 seconds from Jupiter, the easternmost one 30 seconds from the middle one. The western star was 10 minutes from Jupiter. They were nearly in the same straight line, only the eastern one, closest to Jupiter, was a little bit elevated to the north. But at the fourth hour the two to the east were still closer to each

East * * O * West

other, for they were only 20 seconds apart. In these observations the western star appeared very small.

On the first day of February, at the second hour of the night, the formation was similar. The eastern star was 6 minutes from

East * [*]O * West

Jupiter and the western one 8. To the east a very small star was 20 seconds distant from Jupiter. They traced out a precisely straight line.

On the second, the stars appeared in this order. A single star to the east was 6 minutes from Jupiter; Jupiter was 4 minutes distant

East * O * * West

from the nearer one to the west; and between this one and the westernmost star there was an interval of 8 minutes. They were precisely in a straight line and of nearly the same magnitude. But at the seventh hour there were four stars, among which Jupiter

East * * O * * West

occupied the middle position. Of these stars the easternmost one was 4 minutes from the next, this one 1 minute, 40 seconds from Jupiter, Jupiter 6 minutes from the western one closest to him, and this one 8 minutes from the westernmost one. They were all together on the same straight line extended along the line of the zodiac.

On the third, at the seventh hour, the stars were arranged in this sequence. The eastern one was 1 minute, 30 seconds from Jupiter; the closest western one 2 minutes; and the other western one was

East * O * * West

10 minutes removed from this one. They were absolutely on the same straight line and of equal magnitude.

On the fourth, at the second hour, there were four stars around Jupiter, two to the east and two to the west, and arranged precisely

on a straight line, as in the adjoining figure. The easternmost was distant 3 minutes from the next one, while this one was 40 seconds from Jupiter; Jupiter was 4 minutes from the nearest western one, and this one 6 minutes from the westernmost one. Their magnitudes were nearly equal; the one closest to Jupiter appeared a little smaller than the rest. But at the seventh hour the eastern stars were only 30 seconds apart. Jupiter was 2 minutes from the nearer eastern

East ** O * * West

one, while he was 4 minutes from the next western one, and this one was 3 minutes from the westernmost one. They were all equal and extended on the same straight line along the ecliptic.

On the fifth, the sky was cloudy.

On the sixth, only two stars appeared flanking Jupiter, as is seen

East * O * West

in the adjoining figure. The eastern one was 2 minutes and the western one 3 minutes from Jupiter. They were on the same straight line with Jupiter and equal in magnitude.

On the seventh, two stars stood near Jupiter, both to the east, arranged in this manner.

East * * ○ West

The intervals between them and with Jupiter were equal, that is, 1 minute, and a straight line ran through them and the center of Jupiter.

On the eighth, at the first hour, three stars were present, all to the east, as in the figure. The small star closest to Jupiter was 1

East * * * ○ West

minute, 20 seconds distant from him; the middle star was 4 minutes from this one and rather large; and the very small easternmost star was 20 seconds from that one. I was of two minds whether the one closest to Jupiter was only one, or two little stars, for it seemed now and then that there was another star near it, toward the east, extremely small, and separated from it by only 10 seconds. They were all extended on the same straight line along the zodiac. But at the third hour the star closest to Jupiter nearly touched him. It was only 10 seconds from him, while the others had moved farther from Jupiter, for the middle one was 6 minutes away from Jupiter. Finally, at the fourth hour, the one that before was closest to Jupiter, united with him, was seen no longer.

On the ninth, at 30 minutes, two stars were near Jupiter to the east and one to the west, in this formation. The easternmost star,

which was rather small, was 4 minutes from the next one; the larger middle star was 7 minutes distant from Jupiter; Jupiter was 4 minutes removed from the western star, which was small.

On the tenth, at 1 hour, 30 minutes, two very small stars, both to the east, appeared in this arrangement. The farther one was 10

East * * ○ West

minutes from Jupiter and the nearer one 20 seconds. They were on the same straight line. But in the fourth hour the star close to Jupiter did not appear any longer and the other appeared so diminished that it could hardly be perceived, although the air was very clear, and it was farther from Jupiter than it had been before, since it was now 12 minutes distant.

On the eleventh, at the first hour, two stars were present to the east and one to the west. The western one was 4 minutes from

West

Jupiter; the nearer one to the east was likewise 4 minutes away from Jupiter, while the easternmost star was 8 minutes from this one. They were moderately conspicuous and on the same straight line. But at the third hour a fourth star appeared close to Jupiter to the east, smaller than the other ones, separated from Jupiter by 30

East West

seconds and slightly displaced to the north from the straight line drawn through the rest of the stars. They were all most brilliant and very conspicuous. But at the fifth hour plus a half the star closest to Jupiter to the east, already more remote from him, had attained a position in the middle between him and the more eastern star close to itself. And they were all precisely on the same straight line and of the same magnitude, as can be seen in the adjoining figure.

East West

On the twelfth, at 40 minutes, two stars were present to the east and likewise two to the west. The farther one to the east was 10

West

minutes from Jupiter while the more remote star to the west was 8 minutes away. They were both rather conspicuous. The other two stars were very close to Jupiter and very small, especially the eastern one, which was 40 seconds distant from Jupiter, while the western one was 1 minute away. But at the fourth hour the little star that was close to Jupiter to the east no longer appeared.

On the thirteenth, at 30 minutes, two stars appeared to the east and two also to the west. The eastern star closer to Jupiter, fairly

conspicuous, was 2 minutes from him, and the more eastern one, appearing smaller, was 4 minutes removed from this one. The western star farther from Jupiter, exceedingly conspicuous, was separated from him by 4 minutes. Between it and Jupiter fell a small starlet closer to the westernmost star, since it was not more than 30 seconds from it. They were all precisely on the same straight line along the length of the ecliptic.

On the fifteenth (for on the fourteenth the sky was covered by clouds), at the first hour, the position of the stars was as follows.

East * ** O West

That is, there were three stars to the east, but none were seen to the west. The star to the east closest to Jupiter was 50 seconds from him, the next one was 20 seconds from this one, and the easternmost star 2 minutes from this one. And it was larger than the others, for the two nearer ones were exceedingly small. But at about the fifth hour, of the stars close to Jupiter only one was seen, 30 seconds

East * * O West

distant from Jupiter. The elongation of the more eastern one from Jupiter was increased for it was then 4 minutes. But at the sixth hour, in addition to the two positioned to the east, as was stated a

East * *O * West

moment ago, one little star, exceedingly small, was seen toward the west, 2 minutes removed from Jupiter.

On the sixteenth, at the sixth hour, they were in the following arrangement. That is, one star was 7 minutes away from Jupiter to the east, Jupiter 5 minutes from the next star to the west, and this

East * O * * West

one 3 minutes from the remaining western one. They were all of about the same magnitude, fairly conspicuous, and exactly on the same straight line drawn along the zodiac.

On the seventeenth, at the first hour, two stars were present, one to the east 3 minutes from Jupiter and another to the west,

East * O * West

distant 10 minutes. This star was somewhat smaller than the eastern one. But at the sixth hour the eastern one was closer to Jupiter and was only 50 seconds distant from him. The western star was farther, that is, 12 minutes. In both observations they were on the same straight line, and both were rather small, especially the one to the east in the second observation.

On the eighteenth, at the first hour, three stars were present, of which two were to the west and one to the east. The eastern star

East * O * * West

was 3 minutes from Jupiter, the closest one to the west 2 minutes, and the remaining more westerly star was 8 minutes from the middle one. All were precisely on the same straight line and of nearly the same magnitude. But at the second hour the stars closer to Jupiter were removed from Jupiter by equal spaces, for the western one [of these] was now also 3 minutes away from him. But at the sixth hour a fourth star appeared between the eastern one and Jupiter in the following configuration. The easternmost star was 3 minutes from the next one, this star 1 minute, 50 seconds from Jupiter, Jupiter 3 minutes from the next western star, and this one

7 minutes from the westernmost star. They were nearly equal, only the eastern one close to Jupiter was a bit smaller, and they were all on the same straight line parallel to the ecliptic.

On the nineteenth, at 40 minutes, only two stars, rather large,

East ○ * * West

were seen to the west of Jupiter and precisely arrayed with Jupiter on the same straight line drawn along the ecliptic. The nearer star was 7 minutes from Jupiter and 6 minutes from the westernmost star.

On the twentieth, the sky was cloudy.

On the twenty-first, at 1 hour, 30 minutes, three little stars, rather small, were observed in this arrangement. The eastern star

East * ○ * * West

was 2 minutes from Jupiter, Jupiter 3 minutes from the next western one, and this star 7 minutes from the westernmost one. They were precisely in the same straight line, parallel to the ecliptic.

On the twenty-fifth, at 1 hour, 30 minutes (for during the three preceding nights the sky was covered by clouds), three stars ap-

East * * ○ * West

peared, two to the east, whose distances between themselves and from Jupiter were equal at 4 minutes. To the west one star was 2 minutes from Jupiter. They were precisely on the same straight line extending along the ecliptic.

On the twenty-sixth, at 30 minutes, only two stars were present. One was to the east 10 minutes from Jupiter, and the other was to

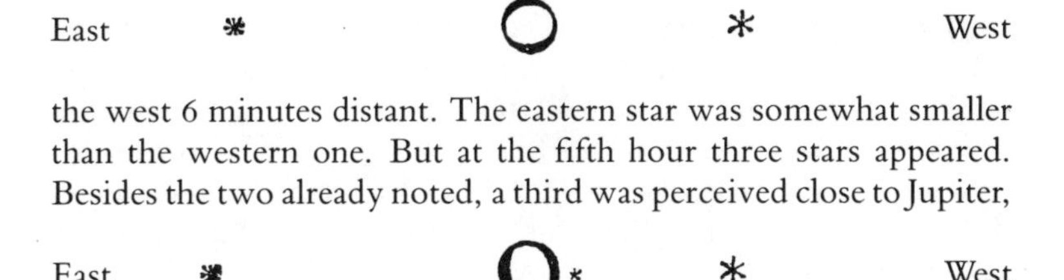

the west 6 minutes distant. The eastern star was somewhat smaller than the western one. But at the fifth hour three stars appeared. Besides the two already noted, a third was perceived close to Jupiter,

to the west and very small, which earlier had been hidden behind Jupiter, and it was 1 minute from him. The eastern star appeared farther than before, being 11 minutes from Jupiter. On this night I decided for the first time to observe the progress of Jupiter and his adjacent planets along the length of the zodiac by reference to some fixed star, for a fixed star was observed to the east, 11 minutes from the easternmost planet and displaced somewhat to the south in the following manner:[89]

East * O* * West

* fixa

On the twenty-seventh, at 1 hour, 4 minutes,[90] the stars appeared in this configuration. The easternmost star was 10 minutes from Jupiter, the next star, close to Jupiter, 30 seconds; the next one, to

East * [*]O * * West

* fixa

89. This is a seventh-magnitude star, just below the ecliptic, at R.A. 5 hours, 4 minutes and decl. +22°.4, in the constellation Taurus.

90. Galileo clearly meant "40 minutes," but this is not a printer's error: the manuscript also has "4 minutes" (*Opere,* 3:44).

the west, was 2 minutes, 30 seconds from Jupiter, and the westernmost star was 1 minute distant from this one. The stars nearer Jupiter appeared small, especially the eastern one, but the outermost stars were very conspicuous, especially the western one. And they formed a straight line exactly drawn along the ecliptic. The progress of these planets toward the east was clearly discerned through a comparison with the aforesaid fixed star, for Jupiter with his attendant planets was closer to it, as can be seen in the adjoining figure. But at the fifth hour the eastern star near Jupiter was 1 minute away from him.

On the twenty-eighth, at the first hour, only two stars were seen, an eastern one 9 minutes, and a western one 2 minutes from Jupiter. They were fairly conspicuous and on the same straight line. This line was perpendicularly intersected by a line from the fixed star to the eastern planet, as shown in the figure.

*fixa

But at the fifth hour a third little star was perceived to the east 2 minutes distant from Jupiter in this arrangement.

East * * O * West

On the first of March, at 40 minutes, four stars were perceived, all to the east. The nearest star to Jupiter was 2 minutes from him, the next star 1 minute from this one, and the third 20 seconds, and

East * * * * O West

*fixa

it was brighter than the rest. The fourth star was 4 minutes from this one, and it was smaller than the rest. They formed nearly a straight line except that the third star from Jupiter was elevated a little. The fixed star formed an equilateral triangle with Jupiter and the easternmost star, as shown in the figure.

On the second, at 40 minutes, three planets were present, two to the east and one to the west, in this configuration.

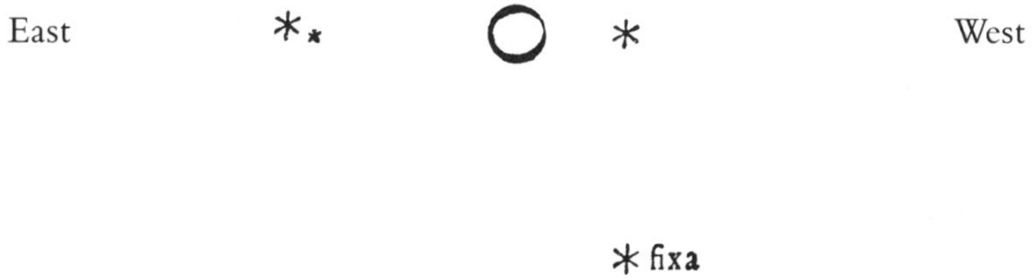

East * * O * West

* fixa

The easternmost planet was 7 minutes from Jupiter, while this one was 30 seconds from the next planet. The western planet was 2 minutes removed from Jupiter. And the outermost planets were brighter and larger than the other one, which appeared very small. The easternmost planet appeared somewhat elevated toward the north above the straight line drawn through Jupiter and the other ones. The fixed star already noted was 8 minutes distant from the western planet along the line drawn to that planet perpendicular to the straight line extended through all the planets, as the figure shows.

I decided to add these comparisons of Jupiter and his adjacent planets with the fixed star so that from them anyone could see that the progress of these planets, in longitude as well as latitude, agrees exactly with the motions that are derived from the tables.[91]

These are the observations of the four Medicean planets recently, and for the first time, discovered by me. From them, although it

91. Jupiter had passed its station at the end of January and was slowly moving from west to east. Its daily motion in longitude was about 4 arcminutes at the end of February. See Bryant Tuckerman, *Planetary, Lunar and Solar Positions A.D. 2 to A.D. 1649 at Five-Day and Ten-Day Intervals,* American Philosophical Society, *Memoirs* 59 (1964): 823.

is not yet possible to calculate their periods, something worthy of notice may at least be said. And first, since they sometimes follow and at other times precede Jupiter by similar intervals, and are removed from him toward the east as well as the west by only very narrow limits, and accompany him equally in retrograde and direct motion, no one can doubt that they complete their revolutions about him while, in the meantime, all together they complete a 12-year period about the center of the world. Moreover, they whirl around in unequal circles, which is clearly deduced from the fact that at the greatest separations from Jupiter two planets could never be seen united while, on the other hand, near Jupiter two, three, and occasionally all four planets are found crowded together at the same time. It is further seen that the revolutions of the planets describing smaller circles around Jupiter are faster.[92] For the stars closer to Jupiter are often seen to the east when the previous day they appeared to the west, and vice versa, while from a careful examination of its previously accurately noted returns, the planet traversing the largest orb appears to have a semimonthly period.[93] We have moreover an excellent and splendid argument for taking away the scruples of those who, while tolerating with equanimity the revolution of the planets around the Sun in the Copernican system, are so disturbed by the attendance of one Moon around the Earth while the two together complete the annual orb around the Sun that they conclude that this constitution of the universe must be overthrown as impossible.[94] For here we have only one planet revolving around another while both run through a great circle around the Sun: but our vision offers us four stars wandering around Jupiter like the Moon around the Earth while all together with Jupiter traverse a great circle around the Sun in the space of

92. Kepler's third law, relating the mean radii of the orbits of the planets to their periods, was not published until 1619.

93. The actual period is about 16 days, 18 hours.

94. This was one of the arguments against the Copernican hypothesis. If the Earth is a planet, why should it be the only planet to have a moon? Alternatively, how could there be two centers of rotation in the universe?

12 years.[95] Finally, we must not neglect the reason why it happens that the Medicean stars, while completing their very small revolutions around Jupiter, are themselves now and then seen twice as large. We can in no way seek the cause in terrestrial vapors, for the stars appear larger and smaller when the sizes of Jupiter and nearby fixed stars are seen completely unchanged. It seems inconceivable, moreover, that they approach and recede from the Earth by such a degree around the perigees and apogees[96] of their orbits as to cause such large changes. For smaller circular motions can in no way be responsible, while an oval motion (which in this case would have to be almost straight) appears to be both inconceivable and by no account harmonious with the appearances.[97] I gladly offer what occurs to me in this matter and submit it to the judgment and censure of right-thinking men. It is well known that because of the interposition of terrestrial vapors the Sun and Moon appear larger but the fixed stars and planets smaller. For this reason, near the horizon the luminaries appear larger[98] but the stars [and planets] smaller and generally inconspicuous, and they are diminished even more if the same vapors are perfused by light. For that reason the stars [and planets] appear very small by day and during twilight, but not the Moon, as we have already stated above.[99] From what we have said above as well as from those things that will be discussed

95. This passage in its entirety removes an important objection against the Copernican theory, for Jupiter's moons demonstrate that our Moon can revolve around a moving Earth. It has been suggested, however, that it is an argument against the geo-heliocentric system of Tycho Brahe. See Wade L. Robison, "Galileo on the Moons of Jupiter," *Annals of Science* 31 (1974): 165–69.

96. Apogee and perigee are the points at which a heavenly body is farthest from, and closest to, the Earth, and Galileo is using the terms here in their literal meanings.

97. Although the orbits of Jupiter's satellites are virtually circles, technically they are ellipses. Elliptical astronomy was introduced by Johannes Kepler in his *Astronomia Nova* of 1609.

98. In fact, atmospheric refraction makes the vertical diameter of these bodies smaller than the horizontal diameter. The large sizes of the Moon and Sun when close to the horizon are optical illusions.

99. See pp. 57–58.

more amply in our system, it is moreover certain that not only the Earth but also the Moon has its surrounding vaporous orb.[100] And we can accordingly make the same judgment about the remaining planets, so that it does not appear inconceivable to put around Jupiter an orb denser than the rest of the ether around which the Medicean planets are led like the Moon around the sphere of the elements. And at apogee, by the interposition of this orb, they are smaller, but when at perigee, because of the absence or attenuation of this orb, they appear larger.[101] Lack of time prevents me from proceeding further. The fair reader may expect more about these matters soon.

100. See note 43, above.

101. The variations in brightness reported by Galileo cannot be accounted for by the varying brightness of individual satellites. Since in Galileo's reports satellites were seen dim only when they were close to Jupiter, this effect must be ascribed to a combination of the glare of the planet and the poor resolution of Galileo's telescope.

CONCLUSION: THE RECEPTION OF SIDEREUS NUNCIUS

Sidereus Nuncius made Galileo into an international celebrity almost overnight. Although news did not travel instantaneously in the seventeenth century as it does today, it traveled surprisingly fast through diplomatic and commercial channels. From Venice it took less than 2 weeks for letters to reach southern Germany, and about a month to reach distant England. And some of the letters transmitted through these channels in the spring of 1610 contained mentions of the astounding discoveries made by the mathematics professor at the University of Padua. Quickly, therefore, Galileo's name was on the lips of learned men all over Europe. The exact nature of the discoveries, however, was usually not known until copies of the book itself arrived in the wake of the rumors. At that point, scholars could read for themselves the claims made by Galileo and the process of evaluation could begin. But few scientists had access to spyglasses other than the ordinary low-powered ones, and even the best instruments could not initially compete for quality with those made by Galileo. High-quality glass was difficult to obtain,[1] while the spectacle-making craft was too tradition-bound to respond quickly to the demand for lenses outside the normal range of strengths.[2] Before scientists could verify or disprove the discoveries they had to procure appropriate instruments, and this took time. Not before the autumn of 1610 was independent verification, in Italy and abroad, forthcoming.

1. Olaf Pedersen, "Sagredo's Optical Researches," *Centaurus* 13 (1968): 139–50; Silvio Bedini, "The Makers of Galileo's Scientific Instruments," in *Atti del simposio internazionale di storia, metodologia, logica e filosofia della scienza "Galileo nella storia e nella filosofia della scienza,"* 4 vols. (Florence: G. Barbera, 1967), 2 (part 5): 89–115.

2. A. O. Prickard, "The 'Mundus Jovialis' of Simon Marius," *Observatory* 39 (1916): 370–71; Girolamo Sirturi, *Telescopium: Sive ars perficiendi* (Frankfurt, 1618), 22–30.

That is not to say, however, that no opinions were expressed before Galileo's claims had been put to the test: controversies about the discoveries flared up very quickly after *Sidereus Nuncius* came off the press.

Galileo's discoveries were controversial for several reasons. First, the spyglass presented methodological and epistemological problems. The prevailing Aristotelian methodology was based on deductions and inferences from information gathered by means of the unaided senses. In some subjects such as anatomy and physiology (which in the sixteenth century became important research areas) Aristotelian methodology was at this very time beginning to bear important fruits. But Galileo's claim was that the spyglass revealed phenomena that were invisible to the naked eye. This was very intriguing, but how could one be certain that the instrument did not deceive Galileo and that the phenomena really did exist in the heavens?

Looking back from our point of view, the problem lay with the role traditionally assigned to mathematics. The relevance of this approach to nature had been severely limited in Aristotelian science. For instance, mathematics could predict where a planet would be in the heavens at a particular time, but it could not inform us on how the heavens were constructed. The models or constructions used by the mathematicians (as astronomers were called) to arrive at their predictions were not regarded as having anything to do with reality: they were mere devices. In other areas of what we might call applied mathematics, the situation was much the same.

Galileo made his living as a mathematician. Like Copernicus before him, he insisted on extending his subject across traditional disciplinary boundaries to make statements about the way the world is actually constructed.[3] According to Galileo, this new optical device—and optics was a branch of mathematics—showed things in the heavens as they really were, even if they were invisible with the naked eye. Methodologically speaking, this was a very bold

3. Robert S. Westman, "The Astronomer's Role in the Sixteenth Century: A Preliminary Study," *History of Science* 18 (1980): 105–47.

claim, for not only was there no optical theory that could demonstrate that the instrument did not deceive the senses, it was not even accepted in principle that optical theory could have much to do with reality. Now, this attitude had been changing in the sixteenth century, but the professors of philosophy were not about to yield this ground to a mathematician without a fight.[4]

We can understand the importance of this issue better when we realize that the telescope was the first scientific instrument that amplified the senses and made hitherto invisible things visible. While today we accept such devices readily—indeed, we believe that they are an essential part of science—the situation was different in 1610. The legitimacy of the telescope was open to question, and so, therefore, was the evidence it presented. It seemed to create entirely new information and did so through optical principles that were not at all well understood. Science had to make room for this new type of instrument and the evidence it produced. And this process was made very difficult by the dearth of good instruments. The moons of Jupiter were the test case of instrument quality, and questions about their existence dominated discussions about the new discoveries. Although this part of the evaluation was over by the spring of 1611, the problem of supplying the telescope with an adequate theory that certified the existence of the phenomena seen through it continued to exercise the scientific community for the rest of the seventeenth century.[5]

Moreover, the evidence supplied by the new instrument flew in the face of the cherished cosmological ideas of the philosophical establishment. Obviously, if the reality of these new phenomena

4. The important philosopher Cesare Cremonini, Galileo's colleague and friend at the University of Padua, wanted nothing to do with the telescope. On 6 May 1611, Paolo Gualdo wrote from Padua to Galileo in Florence that Cremonini "entirely ridicules these observations of yours and is amazed that you assert them as true." See *Opere,* 11:100.

5. As late as 1681, the English astronomer John Flamsteed still deemed it necessary to show by an analysis of lens systems that lenses and combinations of lenses "doe not impose upon our senses." See *The Gresham Lectures of John Flamsteed,* ed. Eric G. Forbes (London: Mansell, 1975), 189.

was accepted, it would be very difficult to fit them into the traditional cosmology and philosophy. One could not maintain the perfection of the heavens if one accepted Galileo's claim that the Moon was covered with mountains and valleys like the Earth. The Copernican hypothesis, which still had few adherents (although their number was growing), could accommodate the discoveries much better, although none constituted proof that this hypothesis was correct. The telescope opened, as it were, a second front in the struggle between the world systems, and the battle now began to intensify.

News of Galileo's celestial discoveries began to spread even before the publication of *Sidereus Nuncius*. On 12 March, the same day that Galileo signed the dedication of his book, Marc Welser, a banker in Augsburg, on the other side of the Alps, wrote to Christopher Clavius, the senior mathematician at the Jesuit Collegio Romano in Rome:[6]

> With this occasion I cannot neglect to tell you that it has been written to me from Padua as a certain and secure thing that with a new instrument called by many *visorio* of which he makes himself the author [*autore*], Mr. Galileo Galilei of that university has discovered four planets, new to us, having never been seen, as far as we know, by a mortal man, and also many fixed stars, not known or seen before, and marvelous things about the Milky Way. I know very well that "to believe slowly is the sinew of wisdom," and I have not made up my mind about anything. I ask Your Reverence, however, freely to tell me your opinion about this fact in confidence.

When *Sidereus Nuncius* itself followed this advance publicity, rulers and church officials, first in Italy and then in other areas of Europe, asked their experts what to think of these claims, and the experts were often at a loss. Some accepted them, others rejected

6. *Opere*, 10:288.

them out of hand, and many reserved their judgment. Europe was already flooded with spyglasses; but ordinary spyglasses, which might only imperfectly show some of the lunar phenomena, could not show the satellites of Jupiter. Recognizing this problem, Galileo set out to convert Europe to his views by sending out powerful spyglasses as well as copies of his book.

As we have seen, in all likelihood Galileo showed Grand Duke Cosimo II what the Moon looked like through an early spyglass during a visit to Florence in the autumn of 1609, and within a week of the publication of *Sidereus Nuncius,* in March 1610, sent him a spyglass (and sent his private secretary, Enea Piccolomini, instructions on its use) through which he could see the Medicean planets.[7] Galileo was aware that even with one of his superior instruments these observations were difficult and it was essential that he convince his Medici patrons of the truth of his discoveries. This was best done by demonstrating the use of his instrument and showing the new phenomena personally, and therefore Galileo went to Tuscany during Easter vacation.[8] By the end of April he had personally seen to it that his discoveries were verified by the Grand Duke. Tuscany was friendly territory for Galileo, but even here the efforts to explain away the discoveries were already beginning. Early in March Raffaello Gualterotti, an old acquaintance, had written to Galileo explaining the spotted nature of the Moon as due to earthly exhalations.[9]

Amid the frantic activity of writing his book, seeing it through the press, and continuing his celestial observations, Galileo had continued making telescopes. Among the many he had made by March 1610, only a few were good enough to show Jupiter's moons.[10] In his letter of 19 March to the Tuscan court, he outlined his plan as follows:[11]

7. Ibid., pp. 299–300.
8. Ibid., pp. 289, 302–3.
9. Ibid., pp. 284–86.
10. Ibid., pp. 298, 302.
11. Ibid., p. 301.

> In order to maintain and increase the renown of these discoveries, it appears to me necessary . . . to have the truth seen and recognized, by means of the effect itself, by as many people as possible. I have done, and am doing, this in Venice and Padua. But spyglasses that are most exquisite and capable of showing all the observations are very rare, and among the sixty that I have made, at great cost and effort, I have been able to find only a very small number. These few, however, I have planned to send to great princes, and in particular to the relatives of the Most Serene Grand Duke. And already I have been asked [for instruments] by the Most Serene Duke of Bavaria and the Elector of Cologne, and also by the Most Illustrious and Reverend Cardinal Del Monte, to whom I shall send [spyglasses] as soon as possible, together with the treatise. My desire would be to send them also to France, Spain, Poland, Austria, Mantua, Modena, Urbino, and wherever else it would please His Most Serene Highness.

It should not surprise us that Galileo sent instruments to rulers and not to scientists, for the men mentioned here, all well intentioned toward Galileo, were patrons of science and would let the instruments be used by their own experts, thus almost guaranteeing a fair hearing.

Among those less well intentioned toward Galileo, things could turn out differently. On his way to Florence in April, Galileo stopped in Bologna, where he visited the internationally renowned astronomer Giovanni Antonio Magini (1555–1617), who was perhaps a bit jealous of his rival's success. A few days later, a young associate of Magini, Martin Horky from Bohemia, wrote the following account to Johannes Kepler (1571–1630), the Imperial Mathematician in Prague:[12]

> Galileo Galilei, the mathematician of Padua, came to us in Bologna and he brought with him that spyglass through which

12. Ibid., p. 343.

> he sees four fictitious planets. On the twenty-fourth and twenty-fifth of April I never slept, day and night, but tested that instrument of Galileo's in innumerable ways, in these lower [earthly] as well as the higher [realms]. On Earth it works miracles; in the heavens it deceives, for other fixed stars appear double. Thus, the following evening I observed with Galileo's spyglass the little star that is seen above the middle one of the three in the tail of the Great Bear, and I saw four very small stars nearby, just as Galileo observed about Jupiter. I have as witnesses most excellent men and most noble doctors, Antonio Roffeni, the most learned mathematician of the University of Bologna, and many others, who with me in a house observed the heavens on the same night of 25 April, with Galileo himself present. But all acknowledged that the instrument deceived. And Galileo became silent, and on the twenty-sixth, a Monday, dejected, he took his leave from Mr. Magini very early in the morning. And he gave no thanks for the favors and the many thoughts, because, full of himself, he hawked a fable. Mr. Magini provided Galileo with distinguished company, both splendid and delightful. Thus the wretched Galileo left Bologna with his spyglass on the twenty-sixth.

In a German sentence at the end of the letter (perhaps so that Magini could not read it), Horky added: "Unknown to anyone, I have made an impression of the spyglass in wax, and when God aids me in returning home, I want to make a much better spyglass than Galileo's." There is no evidence that he ever succeeded in doing so.

Horky's case is extreme. He was ambitious, unscrupulous, and clearly very jealous of Galileo's success. We shall return to him. But the general reception in Bologna is more important. Galileo's log of observations shows that on 25 April he saw two, and the next evening four, satellites of Jupiter.[13] Clearly, even when he was

13. Ibid., 3:436.

there himself with his instrument, it was not easy to convince learned men who were skeptical about his discoveries.

Shortly after his return to Padua a few days later, Galileo received a long letter from Johannes Kepler. Kepler had been an avowed Copernican since his student days and the previous year had published his *Astronomia Nova,* in which he had greatly strengthened the Copernican theory by his demonstration that the orbits of the planets around the Sun were elliptical. Galileo, of course, was very interested in the reaction of Europe's most prestigious astronomer, and he had therefore sent a copy of *Sidereus Nuncius* to the Tuscan ambassador to the Imperial Court at Prague, asking for a written reply by Kepler. The ambassador let Kepler read the book and transmitted Galileo's request. As a result, on 19 April Kepler delivered a long letter on the subject, a letter which he published early in May under the title *Dissertatio cum Nuncio Sidereo,* or "Conversation with the Sidereal Messenger."[14]

Kepler relates how he had first heard the rumor of the four new planets and had correctly surmised that they must be moons. He had read the emperor's copy of *Sidereus Nuncius* even before the ambassador had lent him the book. Spyglasses were readily available in Prague, and Emperor Rudolph II had already observed the Moon through such an instrument at the beginning of 1610, asking Kepler's opinion of lunar spots.[15] But the best instruments in Prague could not show Jupiter's moons, and therefore Kepler had to take this discovery on faith. His statement contrasts sharply with Horky's utterings:[16]

> I may perhaps seem rash in accepting your claims so readily with no support of my own experience. But why should I not believe a most learned mathematician, whose very style attests the soundness of his judgment? He has no intention of

14. Ibid., pp. 97–126. See E. Rosen, *Kepler's Conversation with Galileo's Sidereal Messenger* (New York: Johnson Reprint Corp., 1965).

15. Rosen, *Kepler's Conversation,* 13.

16. Ibid., pp. 12–13.

> practising deception in a bid for vulgar publicity, nor does he pretend to have seen what he has not seen. Because he loves the truth, he does not hesitate to oppose even the most familiar opinions, and to bear the jeers of the crowd with equanimity.

After a discussion of the spyglass (in which he pointed out the defect of spherical aberration and how to prevent it),[17] Kepler turned his attention to Galileo's lunar observations. He had no quarrel with the proposition that, like the Earth, the Moon is rough and uneven. Since the ancient spots, visible with the naked eye, were shown by Galileo to be smoother while the bright areas were dotted with valleys, Kepler pronounced himself convinced that the dark areas must be seas and the bright areas land, as Plutarch had argued 1500 years earlier.[18] But Kepler went further:[19]

> I cannot help wondering about the meaning of that large circular cavity in what I usually call the left corner of the mouth [of the face in the Moon]. Is it a work of nature, or of a trained hand? Suppose there are living beings on the moon . . . [?] It surely stands to reason that the inhabitants express the character of their dwelling place, which has much bigger mountains and valleys than our earth has. Consequently, being endowed with very massive bodies, they also construct gigantic projects. Their day is as long as 15 of our days, and they feel insufferable heat. Perhaps they lack stone for erecting shelters against the sun. On the other hand, maybe they have a soil as sticky as clay. Their usual building plan, accordingly, is as follows. Digging up huge fields, they carry out the earth and heap it in a circle, perhaps for the purpose of drawing out the moisture down below. In this way they may hide in the deep shade behind their excavated mounds and, in keeping with the sun's motion, shift about inside, clinging to the

17. See note 29, p. 14, above. Kepler suggested here a hyperbolic curvature in order to prevent the defect. See Rosen, *Kepler's Conversation,* 19–20.

18. Ibid., pp. 26–27.

19. Ibid., pp. 27–28.

> shadow. They have, as it were, a sort of underground city. They make their homes in numerous caves hewn out of that circular embankment. They place their fields and pastures in the middle, to avoid being forced to go too far away from their farms in their flight from the sun.

In his typical exuberant style, Kepler let his imagination soar. He mentioned that he had entertained such speculations since 1593, and, in fact, his thoughts on this subject were published posthumously as his *Somnium,* or "Dream."[20] Although there were precedents going back to Antiquity, these utterings by Kepler marked the starting point of modern speculations about the possibility of life on other planets.[21]

Galileo had argued that the spyglass strips the adventitious rays away from planets and stars and then magnifies their natural bodies, so that it appeared that the instrument does not magnify these bodies as much as it does the Moon (see pp. 57–58, above). Kepler disagreed with this explanation, arguing that the cause of this phenomenon must be found in the various refractions in the eye itself. The merits of the arguments are perhaps less important than the fact that both Galileo and Kepler had begun to consider the eye itself as an instrument whose strengths and weaknesses could be discussed scientifically: to them the eye itself had become a scientific instrument.[22]

When it came to the difference in appearance between the fixed

20. *Somnium, seu Opus Posthumum de Astronomia Lunari* (1634), reprinted in C. Frisch, ed., *Joannis Kepleri astronomi opera omnia,* 8 vols. (Frankfurt and Erlangen, 1858–71), vol. 8. There are two complete English translations: John Lear, *Kepler's Dream* (Berkeley: University of California Press, 1965); and Edward Rosen, *Kepler's Somnium* (Madison: University of Wisconsin Press, 1967).

21. For the history of ideas about extraterrestrial life, see Steven J. Dick, *Plurality of Worlds: The Origins of the Extraterrestrial Life Debate from Democritus to Kant* (Cambridge: Cambridge University Press, 1982); and Michael J. Crowe, *The Extraterrestrial Life Debate, 1750–1900* (Cambridge: Cambridge University Press, 1986).

22. Harold I. Brown, "Galileo on the Telescope and the Eye," *Journal for the History of Ideas* 46 (1985): 487–501, at 499–501.

stars and the planets, Kepler again drew bold conclusions that went well beyond what Galileo had said:[23]

> What other conclusion shall we draw from this difference, Galileo, than that the fixed stars generate their light from within, whereas the planets, being opaque, are illuminated from without; that is, to use [Giordano] Bruno's terms, the former are suns, the latter, moons or earths?

This was the second time Kepler had invoked the name of Giordano Bruno (ca. 1548–1600), an apostate Dominican friar, who had advocated an infinitude of inhabited worlds and had been burned at the stake in 1600. Kepler had for some years argued vociferously against a universe that is the same everywhere, insisting instead on the privileged position of the Sun and a distinction in kind between the Sun and the fixed stars.[24] He repeated his argument here.[25]

Before leaving the subject of the fixed stars, Kepler expressed his approbation of Galileo's observations and conclusions concerning the Milky Way:[26]

> You have conferred a blessing on astronomers and physicists by revealing the true character of the Milky Way, the nebulae, and the nebulous convolutions. You have upheld those writers who long ago reached the same conclusion as you: they are nothing but a mass of stars, whose luminosities blend on account of the dulness of our eyes.

23. Rosen, *Kepler's Conversation,* 34.

24. Kepler, *De Stella Nova in Pede Serpentarii* (1606), *Johannes Kepler Gesammelte Werke,* vols. 1–10, 13–19 (Munich: C. H. Beck, 1937–), 1:234. See also Alexandre Koyré, *From the Closed World to the Infinite Universe* (Baltimore: Johns Hopkins University Press, 1957; New York: Harper & Row, 1958), 58–76; and Albert Van Helden, *Measuring the Universe: Cosmic Dimensions from Aristarchus to Halley* (Chicago: University of Chicago Press, 1985), 63.

25. Rosen, *Kepler's Conversation,* 34–36.

26. Ibid., p. 36. I have changed the translation slightly.

Kepler reserved his greatest praise for the discovery of Jupiter's satellites. When he had first heard the rumor of the existence of four new planets (before reading *Sidereus Nuncius*), he had feared that perhaps Galileo had found planets around a fixed star. This would have supported the doctrine of Giordano Bruno that Kepler dreaded so much.[27] Having read the book, he was not only reassured, he was overjoyed. Here were four planets the existence of which no one had suspected. This discovery led Kepler to reflect on the progress of science, against the smug philosophers who would not allow anything new under the Sun:[28]

> I have also thought it worth while, in passing, to tweak the ear of the higher philosophy. Let it ponder the question whether the almighty and provident Guardian of the human race permits anything useless and why, like an experienced steward, He opens the inner chambers of his building to us at this particular time. . . . Or does God the creator . . . lead mankind, like some growing youngster gradually approaching maturity, step by step from one stage of knowledge to another? (For example, there was a period when the distinction between the planets and the fixed stars was unknown; it was quite some time before Pythagoras or Parmenides perceived that the evening star and the morning star are the same body [i.e., Venus]; the planets are not mentioned in Moses, Job, or the Psalms.) Let the high philosophy reflect, I repeat, and glance backward to some extent. How far has the knowledge of nature progressed, how much is left, and what may the men of the future expect?

Although in thus stressing the progress of science (a rather novel thought at that time) Kepler might appear to us to be rejecting the purposefulness of the world, this is not his aim. Instead he argues that there must be a purpose for these moons of Jupiter, and that

27. Ibid., pp. 36–39.
28. Ibid., p. 40.

purpose is none other than pleasing the inhabitants of Jupiter who can "behold this wonderfully varied display."[29] After comparing the distances of Jupiter's moons from Jupiter with the distance of our Moon from the Earth, he generalizes the purposes of moons:[30]

> The conclusion is quite clear. Our moon exists for us on the earth, not for the other globes. Those four little moons exist for Jupiter, not for us. Each planet in turn, together with its occupants, is served by its own satellites. From this line of reason we deduce with the highest degree of probability that Jupiter is inhabited.

Furthermore, just as the Earth turns on its axis while the Moon moves about the same axis, Jupiter must turn on the same axis around which its moons revolve.[31] But this type of reasoning brought Kepler perilously close to rejecting anthropocentrism. He retreated quickly, however, and filled several pages with arguments as to why man should be the noblest creature in a universe with other inhabitants.[32] After offering his own speculation on why Jupiter's moons might vary in brightness, Kepler ended this tract with a plea for Galileo to carry on his important observations.

Kepler's *Conversation* was a very different work from *Sidereus Nuncius*. Whereas Galileo reported observations and drew some cautious conclusions from them, Kepler, as was his wont, gave his imagination free rein and speculated widely about the meaning of Galileo's discoveries. If some of Kepler's utterings did perhaps not help Galileo's case very much, his ungrudging acceptance of the discoveries, even though he was unable to verify them, had the effect of throwing the prestige of the Imperial Mathematician behind Galileo's observations. The *Conversation* was reprinted in Florence later in 1610.

29. Ibid.
30. Ibid., p. 42.
31. Ibid.
32. Ibid., pp. 43–46.

In the meantime, Galileo was busy defending his discoveries at home. He gave three public lectures at Padua, during which, as he claimed, he managed to convince his most bitter opponents of the truth of his discoveries.[33] He also received many letters in which objections to his discoveries were put forward, and answering them all was a frustrating business:[34]

> It is true that their reasons for mistrust are very frivolous and childish, since they persuade themselves that I am so rash that in testing my instrument a hundred thousand times on a hundred thousand stars and other objects I have not known, or been able to recognize, those deceptions that they think they have recognized without ever having seen the instrument; or else, that I am so stupid that without any need I have wished to compromise my reputation and to ridicule my Prince. The spyglass is very truthful, and the Medicean planets are planets and, like the other planets, will always be.

Galileo did not have to fight this battle alone, however. After his Easter visit to Tuscany, Grand Duke Cosimo II was convinced of the reality of the discoveries. He had decided to give Galileo a position at the Tuscan court,[35] and the Medicean planets now became a matter of state. The Tuscan ambassadors at the courts in Prague, London, Paris, and Madrid were alerted that Galileo would send them copies of his book and perhaps spyglasses as well, and they were instructed to use their good offices to promote Galileo's discoveries. The expenses that Galileo incurred in making the necessary spyglasses were borne by the Tuscan treasury.[36] This was powerful help indeed.

In June there appeared in Modena a little tract written by Martin Horky, who had been present at Galileo's visit to Bologna. The tract, entitled *Brevissima peregrinatio contra nuncium sidereum nuper ad*

33. *Opere,* 10:349.
34. Ibid., p. 357.
35. Ibid., pp. 350–53, 355–56,
36. Ibid., p. 356.

omnes philosophos et mathematicos emissum a Galileo Galilaeo ("A Very Brief Journey against the Sidereal Message Recently Sent to All Philosophers and Mathematicians by Galileo Galilei"), was dedicated to the faculty of the University of Bologna. Horky concerned himself only with Jupiter's moons, arguing that they could not be seen in Bologna when Galileo tried to demonstrate them because they simply did not exist. But Horky's arguments were specious, and the tract was little more than a personal attack on Galileo. For example, referring to some recently published hoaxes, he castigated Galileo as follows:[37]

> If Thomas Narrenhandler knows how to square the circle; if Klappus knows how to make the Philosopher's Stone; if Keknasel announces that he has found the duplication of the cube; then also the sidereal messenger can display and defend new planets around Jupiter.

The scurrility of this attack outraged the skeptical Magini, and upon Horky's return from Modena Magini literally chased him out of his house.[38] Kepler likewise broke all relations with Horky.[39] Horky was publicly put in his place by Galileo's student John Wodderburn,[40] while the reputation of the University of Bologna was rescued by the philosopher Antonio Roffeni (1580–1643).[41] But the low nature of Horky's attack must not blind us to the two great

37. Ibid., 3:139. Squaring the circle (that is, finding the relationship between the circumference and diameter) and duplicating the cube (that is, constructing a cube having twice the volume of a given cube) were problems that had exercised mathematicians since Antiquity. The search for the Philosopher's Stone had been a quest of some alchemists since about the beginning of the Christian era. Narrenhandler, Klappus, and Keknasel are made-up names.

38. *Opere,* 10:375–76.

39. Ibid., pp. 414–17, 419.

40. *Quatuor problematum quae Martinus Horky contra Nuntium Sidereum de quatuor planetis novis disputanda proposuit* (Padua, 1610). See *Opere,* 3:147–78.

41. *Epistola apologetica contra caecam peregrinationem cuiusdam furiosi Martini, cognomine Horkij editam adversus nuntium sidereum* (Bologna, 1610). See *Opere,* 3:191–200.

problems, the optical and the philosophical, *Sidereus Nuncius* posed for the intellectual community.

When at the end of May Jupiter disappeared in the rays of the Sun,[42] Galileo was the only observer who had made a reasonably continuous series of observations of Jupiter's moons. It appears that only those to whom he had personally shown the moons had seen them at all. Independent verification from outside the Galileo camp was still not forthcoming. Galileo put off any thought of reprinting *Sidereus Nuncius* until the reappearance of Jupiter, so that he could incorporate a longer series of observations. He also wished to include better illustrations of the Moon and to reply to all the doubts and difficulties that had been posed.[43] But nothing came from this intention: no second edition of *Sidereus Nuncius* was ever prepared by Galileo. Later in 1610, however, an unauthorized reprint with inferior illustrations was issued in Frankfurt.

Jupiter became visible again, in the morning sky, at the end of July.[44] That month also Galileo made another discovery. The planet Saturn, whose image is much smaller than Jupiter's because of its greater distance, was coming into favorable position for observation, and Galileo discovered that this planet does not have a simple round shape as Jupiter does. He wrote to the Grand Duke's secretary:[45]

> [T]he star of Saturn is not a single one, but an arrangement of three that almost touch each other and never move or change with respect to each other; and they are placed on a line along the zodiac, the one in the middle being about three times larger than the other two on the sides; and they are situated in this form oOo [.]

42. Galileo recorded his last observation before conjunction on 21 May 1610. See *Opere,* 3:437.

43. Ibid., 10:373.

44. Galileo made his first observations of the satellites on 25 July. See *Opere,* 3:439.

45. Ibid., 10:410.

This was the beginning of an astronomical enigma. Saturn appeared to have two companions, but they were very different from the companions of Jupiter. In Saturn's case they were very large, almost touched the planet, and never moved with respect to it. Galileo's best telescopes—and he had made further improvements since January—were not good enough to show Saturn's ring, which was at this time very narrow. The problem posed by Saturn's appearances was to take almost half a century to solve.[46]

For the time being, Galileo wanted this new discovery kept secret, so that he could announce it in the planned new edition of *Sidereus Nuncius.*[47] In the meantime he hid the discovery in an anagram, *smaismrmilmepoetaleumibunenugttauiras.* He could thus announce that he had made a new discovery without revealing its exact nature, thereby eliminating the possibility of someone falsely claiming that he had discovered it before Galileo. In the days before scientific papers, this was a fairly effective device for safeguarding priority, and Galileo was to use it again. His successors also had occasional recourse to it.

Galileo sent this anagram to correspondents including the Jesuit fathers at the Collegio Romano in Rome and Kepler in Prague.[48] Kepler's response is most interesting. He naturally supposed that Galileo had made a discovery about one of the other planets. Since the Earth has one moon and Jupiter had now been shown to have four, Kepler speculated that Mars must have two moons.[49] This speculation was taken up by others, most notably Jonathan Swift in the eighteenth century,[50] and was ultimately proved correct in 1877 when, with a telescope hundreds of times more powerful than

46. Albert Van Helden, "Saturn and His Anses," *Journal for the History of Astronomy* 5 (1974): 105–21; and " 'Annulo Cingitur': The Solution of the Problem of Saturn," ibid. 5 (1974): 155–74.

47. *Opere,* 10:410.

48. Ibid., 19:611.

49. Ibid., 3:185.

50. In "A Voyage to Laputa," Gulliver marvels at the telescopes used by the Laputan astronomers: "They have made a Catalogue of ten Thousand fixed Stars,

Galileo's, Asaph Hall discovered Mars's two tiny moons, Phobos and Deimos.[51]

By September Galileo had come to Florence to take up his new position as principal mathematician of the University of Pisa (a position without duties) and mathematician and philosopher of the Grand Duke.[52] Because of the move, he was only able to make about a dozen observations of Jupiter's satellites in the morning sky before November, when he finally settled into a house.[53] It was during this period that, finally, independent verification of the satellites of Jupiter was forthcoming. Late in September he received word that his friend Antonio Santini in Venice had seen all four satellites on consecutive mornings,[54] and at about the same time he heard that Johannes Kepler in Prague had been able to observe them as well.[55] Kepler had been able to use the telescope Galileo had sent earlier that year to the Elector of Cologne, and he had observed the satellites from 30 August to 9 September.[56] He sent Galileo a little tract entitled *Ioannis Kepleri . . . Narratio de observatis a se quatuor Iovis satellitibus erronibus,* or "Johannes Kepler's Narration about

whereas the largest of ours do not contain above one third Part of that Number. They have likewise discovered two lesser Stars, or *Satellites,* which revolve about *Mars;* whereof the innermost is distant from the Center of the primary Planet exactly three of his Diameters, and the outermost five; the former revolves in the Space of ten Hours, and the latter in Twenty-one and an Half; so that the Squares of their periodical Times, are very near in the same Proportion with the Cubes of their Distances from the Center of *Mars;* which evidently shews them to be governed by the same Law of Gravitation, that influences the other heavenly Bodies." See *The Prose Works of Jonathan Swift,* ed. Herbert Davis, 14 vols. (Oxford: Basil Blackwell, 1939–68), vol. 11: *Gulliver's Travels,* 154–55.

51. Owen Gingerich, "The Satellites of Mars: Prediction and Discovery," *Journal for the History of Astronomy* 1 (1970): 109–15.

52. *Opere,* 10:400, 429.

53. Ibid., 3:439.

54. Ibid., 10:435, 437.

55. Ibid., pp. 436, 439–40.

56. Ibid., 3:184–87.

Four Wandering Companions of Jupiter Observed by Him."[57] Unknown to Galileo, Jupiter's satellites were observed in England by Thomas Harriot beginning on 27 October,[58] and in Aix-en-Provence, in the south of France, by Joseph Gaultier de La Valette (1564–1647) and Nicolas-Claude Fabri de Peiresc (1580–1638) beginning on 24 November.[59] Much later, in 1614, the German astronomer Simon Marius claimed to have discovered Jupiter's moons in December 1609 and had begun recording his observations on 8 January 1610.[60] Marius's controversial claim has been debated for almost four centuries now,[61] but since it is irrelevant to our story we will ignore it.

By the autumn of 1610, then, the existence of Jupiter's moons had been verified, and with this verification came a large measure of certification of the instrument: the spyglass did not deceive one's sight when pointed to the heavens. It was a device that could not be ignored and that was to change astronomy forever. And as if to confirm this point, Galileo announced a major new discovery that

57. Ibid., pp. 181–90. There is some confusion about the date of publication of this tract. The title page indicates 1611, but Kepler sent Galileo a copy with his letter of 25 October 1610 (*Opere* 10:457). See Max Caspar, *Bibliographia Kepleriana* (Munich: C. H. Beck, 1936), 60; reprint (Munich: C. H. Beck, 1968), 52–53. The tract was reprinted in Florence in 1611.

58. Harriot Papers, Petworth MSS HMC 241/4, f. 3. Note that Harriot's dates are Julian. See also John Roche, "Harriot, Galileo, and Jupiter's Satellites," *Archives internationales d'histoire des sciences* 32 (1982): 9–51.

59. Pierre Humbert, "Joseph Gaultier de La Valette, astronome provençal (1564–1647)," *Revue d'histoire des sciences et de leurs applications* 1 (1948): 316.

60. Simon Marius, *Mundus Iovialis* (Nuremberg, 1614). See "The 'Mundus Jovialis' of Simon Marius," tr. A. O. Prickard, *Observatory* 39 (1916): 371–72. Note that Marius's dates are Julian.

61. See, for example, Joseph Klug, "Simon Marius aus Gunzenhausen und Galileo Galilei," *Abhandlungen der II. Klasse der Königliche Akademie der Wissenschaften,* 22:385–526; J. A. C. Oudemans and J. Bosscha, "Galilée et Marius," *Archives Néerlandaises des Sciences Exactes et Naturelles, publiées par la Société hollandaise des Sciences,* series 2 (1903), 8:115–89; J. Bosscha, "Simon Marius, réhabilitation d'un astronome calomnié," ibid. (1907), 12:258–307, 490–527.

bore directly on the growing Copernican controversy. From Copernicus's discussion of the order of the planets in book I, chapter 10, of *De Revolutionibus,* it was clear that Venus ought to appear much larger at perigee than at apogee, and if it shines with borrowed light, it ought to exhibit a full range of phases, just as the Moon does.[62] At its greatest elongation from the Sun, Venus is the brightest object in the heavens. This great brightness, however, presented a problem to Galileo and his successors in the seventeenth century. When seen through telescopes that are not corrected for chromatic aberration, the planet presents a confused image surrounded by color fringes. Its true shape was therefore difficult to make out. When Galileo was making his first discoveries with the spyglass, Jupiter was in the evening sky while Venus was in the morning sky. No doubt Venus's great brightness and the imperfections of his early spyglasses defeated him. But after conjunction with the Sun, in 1610, the planet became visible in the evening sky in October, and Galileo now made a determined assault on the planet with improved instruments. On 5 December Benedetto Castelli, a former pupil of Galileo now living in Brescia, where he had no access to a good spyglass, wrote to Galileo:[63]

> Since (as I believe) the opinion of Copernicus that Venus revolves about the Sun is correct, it is clear that she would necessarily be seen by us sometimes horned and sometimes not, although the said planet is at equal distances from the Sun, at those times, that is, when the smallness of the horns and the effusion of rays do not impede the observation of this difference. Now I would like to know from you if you, with your wonderful spyglasses, have noticed such an appearance, which, without doubt, will be a sure means of convincing any obstinate mind.

62. *Copernicus: On the Revolutions of the Heavenly Spheres,* tr. A. M. Duncan (Newton Abbot: David & Charles; New York: Barnes & Noble, 1976), 47–48. For earlier conjectures about the phases of Venus, see Roger Ariew, "The Phases of Venus before 1610," *Studies in the History and Philosophy of Science* 18 (1987): 81–92.

63. *Opere,* 10:481–82.

As Castelli was writing this letter, Galileo was almost ready to announce his findings. By the beginning of December Venus's disk had been reduced to a little half-moon, and Galileo was now reasonably certain that a further reduction would occur, giving the planet a crescent, or horned, appearance.[64] On 11 December, therefore, he wrote to Giuliano de' Medici, the Tuscan ambassador in Prague, that he had observed a phenomenon that was a strong argument for the Copernican theory.[65] He hid the discovery in an anagram, *Haec immatura a me iam frustra leguntur o y.*[66] When by the end of the month Galileo had actually seen Venus's disk shrink to less than a half-moon and become horned, he could confidently reveal his new discovery. The solution to the anagram was *Cynthiae figuras aemulatur mater amorum,* that is, "the mother of love [Venus] emulates the figures of Cynthia [the Moon]."[67] In other words, Venus goes through phases just as the Moon does. Galileo described to Castelli what he had seen:[68]

> Know, therefore, that about 3 months ago I began to observe Venus with the instrument, and I saw her in a round shape and very small. Day by day she increased in size and maintained that round shape until finally, attaining a very great distance from the Sun, the roundness of her eastern part began to diminish, and in a few days she was reduced to a semicircle. She maintained this shape for many days, all the while, however, growing in size. At present she is becoming

64. Richard S. Westfall has charged that Galileo in fact did not observe Venus until Castelli drew his attention to the planet in December. See "Science and Patronage: Galileo and the Telescope," *Isis* 76 (1985): 11–30. For replies to Westfall, see Stillman Drake, "Galileo, Kepler, and the Phases of Venus," *Journal for the History of Astronomy* 15 (1984): 198–208; Owen Gingerich, "Phases of Venus in 1610," ibid., pp. 209–10; William T. Peters, "The Appearances of Venus and Mars in 1610," ibid., pp. 211–14.

65. *Opere,* 10:483.

66. The sentence can be translated as "These unripe matters are brought together by me in vain."

67. *Opere,* 11:12.

68. Ibid., 10:503.

sickle-shaped, and as long as she is observed in the evening her little horns will continue to become thinner, until she vanishes. But when she then reappears in the morning, she will appear with very thin horns, again turned away from the Sun, and will grow to a semicircle at her greatest digression [from the Sun]. She will then remain semicircular for several days, although diminishing in size, after which in a few days she will progress to a full circle. Then for many months she will appear, both in the morning and then in the evening, completely circular but very small in size.

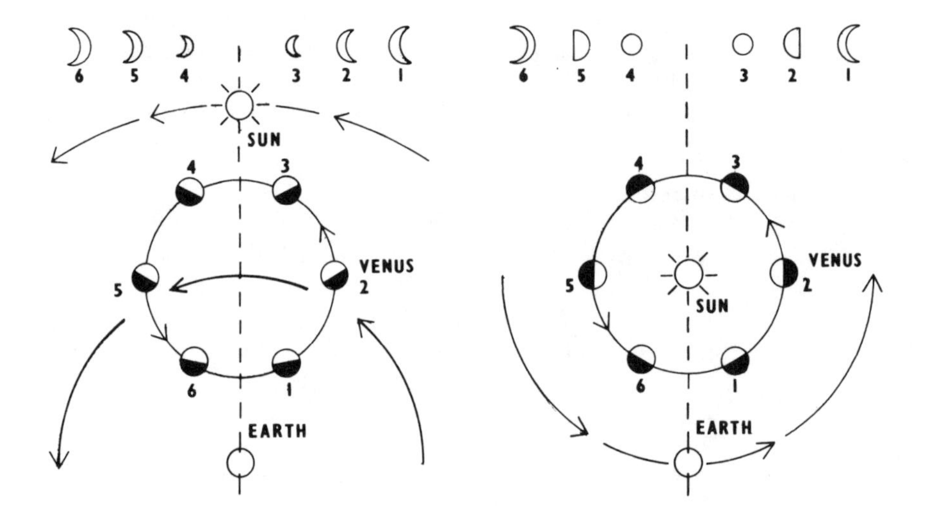

The appearance of Venus predicted by the
Ptolemaic and Copernican systems

Now, this progression of phases proved several things. First, Venus shines with light borrowed from the Sun, just as our Moon does. Second, Venus (and by implication Mercury) goes around the Sun. In the Ptolemaic system, the order of the planets was a convention: in most versions Venus was "below" the Sun; in some versions it was "above" it. But regardless of what one thought the order of the planets was in the Ptolemaic system, Venus was either always between the Sun and the Earth or always beyond the Sun, and neither choice could account for the observed phases (see figure). Galileo's observations therefore proved the Ptolemaic scheme in error. Only a hybrid Ptolemaic system in which Venus and Mercury go around the Sun but all other planets go around the Earth, the Tychonic system in which the Moon and Sun go around the Earth but all other bodies go around the Sun, or the Copernican system could account for Venus's observed phases. Galileo was greatly strengthened in his Copernicanism by these observations.

By the end of 1610, then, the spyglass was established as an instrument that could not be ignored. The satellites of Jupiter had now been verified by a number of independent observers, and Kepler had even published a tract about his observations of the satellites, his *Narratio*. Father Clavius, the chief mathematician at the Collegio Romano, had written Galileo that he had seen innumerable stars invisible with the naked eye, that he had satisfied himself that the bodies moving around Jupiter were indeed moons of that planet, and that he had noticed that Saturn has not a round but an oval shape.[69] These verdicts, coming from the center of astronomical orthodoxy, were welcome news indeed for Galileo,[70] and it was now clear that he was winning the battle for acceptance of the new instrument. His final triumph came a few months later in Rome. Galileo had wanted to visit that city for some time,[71] but poor health prevented him from doing so until the end of March

69. Ibid., pp. 484–85.

70. Note that Clavius had been very skeptical about Galileo's discoveries. In a letter of 1 October 1610, Galileo's friend Lodovico Cardi da Cigoli reported from Rome that "Clavius said to one of my friends about the four stars [revolving around Jupiter] that he laughs at them, and that it will be necessary to make a spyglass which produces them and then shows them, and that Galileo can keep his opinion and he will keep his." See *Opere,* 10:442.

71. *Opere,* 10:432.

1611. His visit to the eternal city was a celebration, in which two events in particular stand out.

The news of the new instrument and its discoveries had, of course, not gone unnoticed by church officials. The implications of the discoveries were not lost on the men who concerned themselves with matters of religious orthodoxy, for Aristotelian philosophy and Christian theology were closely intertwined. Just prior to Galileo's visit, on 19 March 1611, Cardinal Robert Bellarmine, S.J. (1542–1621), head of the Collegio Romano, wrote to his Jesuit mathematicians as follows:[72]

> I know that your Reverences have heard about the new astronomical observations by an eminent mathematician by means of an instrument called a tube or glass; and even I have seen, by means of the same instrument, some very wonderful things concerning the Moon and Venus. I wish therefore that you would favor me with your honest opinion on the following matters:
>
> First, whether you confirm the multitude of fixed stars invisible with the naked eye, and in particular that the Milky Way and the nebulosities are a congeries of very small stars.
>
> 2. That Saturn is not a simple star but three stars joined together.
>
> 3. That the star of Venus changes its shape, waxing and waning like the Moon.
>
> 4. That the Moon has a rough and uneven surface.
>
> 5. That about four movable stars go around the planet of Jupiter, and with motions different among themselves and very swift.
>
> This I wish to know because I hear different opinions, and you Reverend Fathers, being skilled in the mathematical sciences, can easily tell me if these new discoveries are well founded, or if they are apparent and not real.

72. Ibid., 11:87–88. See also James Broderick, *Robert Bellarmine: Saint and Scholar* (Westminster, MD: Newman Press, 1961), 343.

The four mathematicians, Fathers Clavius, Grienberger, Lembo, and Maelcote, replied as follows 5 days later:[73]

> As to the first, it is true that with the spyglass very many wonderful stars appear in the nebulosities of Cancer and the Pleiades, but of the Milky Way it is not so certain that all of it consists of minute stars, and it appears more probable that there are continuous denser parts, although it cannot be denied that in the Milky Way, too, there are many minute stars. It is true that from what is seen in the nebulosities of Cancer and the Pleiades it can be conjectured as probable that in the Milky Way, too, there are a vast multitude of stars that cannot be discerned because they are too small.
>
> As to the second, we have observed that Saturn is not circular, as Jupiter and Mars appear, but is of an oval figure, like this oOo , although we have not seen the two starlets on either side sufficiently separated from the one in the middle to be able to say that they are distinct stars.
>
> As to the third, it is very true that Venus wanes and waxes like the Moon. And having seen her almost full when she was an evening star, we have observed that the illuminated part, which was always turned toward the Sun, decreased little by little, becoming ever more horned. And observing her then as a morning star, after conjunction with the Sun, we saw her horned with the illuminated part toward the Sun. And now the illuminated part continuously increases, according to the light, while the apparent diameter decreases.
>
> As to the fourth, the great inequality of the Moon cannot be denied. But it appears to Father Clavius more probable that the surface is not uneven, but rather that the lunar body is not of uniform density and has denser and rarer parts, as are the ordinary spots seen with the natural sight. Others think that the surface is indeed uneven, but thus far we are not certain enough about this to confirm it indubitably.

73. *Opere,* 11:92–93.

> As to the fifth, four stars go about Jupiter, which move very swiftly, now all to the east, and now all to the west, and sometimes some move to the east and some to the west, all in an almost straight line. These cannot be fixed stars, for they have very swift motions, very different from those of the fixed stars, and they always change their distances from each other and from Jupiter.

As Clavius's opinions about the Moon show, the mathematicians did not necessarily agree with Galileo's interpretations of the observations. The issue at hand, however, was the question posed by Bellarmine, Are these discoveries real or apparent, or, in other words, does the spyglass show things the way they really are or does it deceive the senses? In their response the mathematicians were unanimous: the discoveries were real; the spyglass does not deceive us. Thus the Catholic Church's own expert mathematicians, men of unimpeachable orthodoxy, had now certified the spyglass as a genuine scientific instrument. The mathematicians went so far as to honor Galileo at the Collegio Romano, an occasion at which Maelcote delivered an address in which he expressed his agreement with Galileo's discoveries.[74]

During his stay in Rome, Galileo showed his discoveries through his own spyglass to a large number of influential men. He was feted on many occasions. One of these feasts was given by Federico Cesi (1585–1630), marquis of Monticello, the founder of a scientific academy, the *Accademia dei Lyncei* ("Academy of the Lynx-eyed"). In 1611 it had five members. At the banquet in honor of Galileo, on 14 April, the Florentine was made the sixth member of the Academy.[75] It was during this banquet that the name *telescopium* was given to Galileo's instrument. The term is probably the invention of the Greek poet and theologian John Desmiani (d. 1619).[76]

74. Ibid., 3:291–98.

75. Edward Rosen, *The Naming of the Telescope* (New York: Henry Schuman, 1947), 31. Rosen erred in the number of members of the Academy.

76. Ibid., passim.

Galileo's visit to Rome marked the end of the arguments about the new instrument and the reality of the phenomena it revealed. Although some persisted in their refusal to accept the new instrument, they quickly became isolated. The telescope's validity as an instrument had been amply demonstrated. Henceforth the arguments revolved around the interpretation of the phenomena it had revealed. While in Rome, Galileo had shown spots on the Sun to several observers, and in 1612 he became embroiled in a controversy about the nature of these spots with the German Jesuit Christoph Scheiner (1573–1650). In his effort to save the hallowed idea of the Sun's perfection, Scheiner argued that the spots were not on the Sun but were caused by swarms of satellites! Satellites, those bodies so controversial in 1610, had become commonplace by 1612.

Galileo's troubles were by no means over in 1611. Now that the telescope was becoming an accepted instrument and its revelations could no longer be ignored, the implications of those revelations became pressing. The course of the struggle between geocentric and heliocentric cosmologies was changed irrevocably by the telescope. The instrument forced scientists to reconsider their most fundamental philosophical and cosmological assumptions. Even those who clung tenaciously to the old geocentric cosmology were deeply affected. If the heliocentric camp won in the end, the battle was fierce and Galileo became its most celebrated casualty.

BIBLIOGRAPHY

Adams, C. W. "A Note on Galileo's Determination of the Height of Lunar Mountains." *Isis* 17 (1932): 427–29.

Ambassades du Roy de Siam envoyé à l'Excellence du Prince Maurice arrivé à la Haye le 10. Septemb. 1608. The Hague, 1608. Reprinted in Stillman Drake, *The Unsung Journalist and the Origin of the Telescope.* Los Angeles: Zeitlin & Ver Brugge, 1976.

Ariew, Roger. "Galileo's Lunar Observations in the Context of Medieval Lunar Theory." *Studies in the History and Philosophy of Science* 15 (1984): 213–26.

———. "The Phases of Venus before 1610." *Studies in the History and Philosophy of Science* 18 (1987): 81–92.

Bacon, Roger. *The Opus Maius of Roger Bacon.* Translated by Robert Belle Burke. 2 vols. Philadelphia: University of Pennsylvania Press, 1928.

Bedini, Silvio A. "The Makers of Galileo's Instruments." In *Atti del simposio internazionale di storia, metodologia, logica, e filosofia della scienza "Galileo nella storia e nella filosofia della scienza,"* 4 vols. 2 (part 5): 89–115. Florence: G. Barbera, 1967.

Bloom, Terrie. "Borrowed Perceptions: Harriot's Maps of the Moon." *Journal for the History of Astronomy* 9 (1978): 117–22.

Blumenberg, Hans. *Galileo Galilei Sidereus Nuncius Nachricht von neuen Sternen. Dialog über die Weltsysteme (Auswahl). Vermessung der Hölle Dantes. Marginalien zu Tasso. Herausgegeben und eingeleitet von Hans Blumenberg.* Frankfurt am Main: Insel Verlag, 1965. *Sidereus Nuncius* (pp. 79–131) is translated by Malte Hossenfelder.

Bonelli, Maria Luisa Righini, and William Shea, eds. *Reason, Experiment, and Mysticism in the Scientific Revolution.* New York: Science History Publications, 1975.

Bosscha, J. "Simon Marius, réhabilitation d'un Astronome Calomnié." *Archives Néerlandaises des Sciences Exactes et Naturelles,* series 2, 12 (1907): 258–307, 490–527.

Bosscha, J., and J. A. C. Oudemans. "Galilée et Marius." *Archives Néerlandaises des Sciences Exactes et Naturelles,* series 2, 8 (1903): 115–89.

Broderick, James. *Robert Bellarmine: Saint and Scholar.* Westminster, MD: Newman Press, 1961.

Brown, Harold I. "Galileo on the Telescope and the Eye." *Journal for the History of Ideas* 46 (1985): 487–501.

Busnelli, Manlio Duilio. "Un Carteggio Inedito di Fra Paolo Sarpi con l'Ugonotto Francesco Castrino." *Atti del Reale Istituto Veneto di Scienze, Lettere ed Arti* 87, part 2 (1927–28): 1025–1163.

———, ed. *Fra Paolo Sarpi, Lettere ai Protestanti.* 2 vols. Bari: Gius. Laterza & Figli, 1931.

Cajori, Florian. "History of Determinations of the Heights of Mountains." *Isis* 12 (1929): 482–514.

Campani, Giuseppe. *Ragguaglio di due Nuove Osservazioni.* Rome, 1664.

Cardini, Maria Timpanaro. *Galileo Galilei Sidereus Nuncius. Traduzione con Testo a Fronte e Note di Maria Timpanaro Cardini.* Florence: Sansoni, 1948.

Carlos, Edward Stafford. *The Sidereal Messenger of Galileo Galilei and a Part of Kepler's Dioptrics containing the original account of Galileo's astronomical discoveries. A translation with introduction and notes by Edward Stafford Carlos.* London, 1880. Reprinted, London: Dawsons of Pall Mall, 1960.

Caspar, Max. *Bibliographia Kepleriana.* Munich: C. H. Beck, 1936. 2d ed. edited by Martha List. Munich: C. H. Beck, 1968.

Castelet, Abbé de. See Tinelis, Alexandre.

Chapin, Seymour L. "The Astronomical Activities of Nicolas Claude Fabri de Peiresc." *Isis* 48 (1957): 13–29.

Chitt, José Fernandes. *El Mensajero de los Astros.* Translated by José Fernandes Chitt. Introduction by José Babini. Buenos Aires: Editorial Universitaria de Buenos Aires, 1964.

Clavius, Christophorus. *In Sphaeram Ioannis de Sacro Bosco Commentarius.* Rome, 1570.

Copernicus, Nicholas. *Copernicus: On the Revolutions of the Heavenly Spheres.* Translated by A. M. Duncan. Newton Abbot: David & Charles; New York: Barnes & Noble, 1976.

———. *Nicholas Copernicus on the Revolutions.* Translated by Edward Rosen. Vol. 2, *Nicholas Copernicus Complete Works.* Warsaw and Cracow: Polish Scientific Publishers; London: Macmillan, 1972–85. Published separately, Baltimore: Johns Hopkins University Press, 1978.

Crowe, Michael J. *The Extraterrestrial Life Debate, 1750–1900.* Cambridge and New York: Cambridge University Press, 1986.

Da Vinci, Leonardo. Codex Leicester-Hammer. See Roberts, Jane.

Dick, Steven J. *Plurality of Worlds: The Origins of the Extraterrestrial Life Debate from Democritus to Kant*. Cambridge and New York: Cambridge University Press, 1982.

Drake, Stillman. *Discoveries and Opinions of Galileo, translated with an introduction and notes by Stillman Drake*. Garden City, NY: Doubleday & Co., 1957.

———. "Galileo, Kepler, and the Phases of Venus." *Journal for the History of Astronomy* 15 (1984): 198–208.

———. *Galileo at Work: His Scientific Biography*. Chicago: University of Chicago Press, 1978.

———. "Galileo's First Telescopic Observations." *Journal for the History of Astronomy* 7 (1976): 153–68.

———. Galileo's Steps to Full Copernicanism and Back." *Studies in the History and Philosophy of Science* 18 (1987): 93–105.

———. "The Starry Messenger." *Isis* 49 (1958): 346–47.

———. *Telescopes, Tides, and Tactics*. Chicago: University of Chicago Press, 1983.

———. *The Unsung Journalist and the Origin of the Telescope*. Los Angeles: Zeitlin & Ver Brugge, 1976.

Drake, Stillman, and C. D. O'Malley, eds., trans. *The Controversy on the Comets of 1618*. Philadelphia: University of Pennsylvania Press, 1960.

Favaro, Antonio. *Galileo Galilei e lo Studio di Padova*. 2 vols. Padua, 1883. 2d ed., Padua: Antenore, 1966.

———, ed. *Le Opere di Galileo Galilei*. Edizione Nazionale. 20 vols. Florence: G. Barbera, 1890–1909; reprinted 1929–39, 1964–66.

Flamsteed, John. *The Gresham Lectures of John Flamsteed*. Edited by Eric G. Forbes. London: Mansell, 1975.

Galilei, Galileo. *Dialogue concerning the Two Chief World Systems—Ptolemaic and Copernican*. Translated by Stillman Drake. Berkeley: University of California Press, 1967.

———. *Discourse on Bodies in Water*. Translated by Thomas Salusbury. Edited by Stillman Drake. Urbana: University of Illinois Press, 1960.

———. *Discoveries and Opinions of Galileo, translated with an introduction and notes by Stillman Drake*. Garden City, NY: Doubleday & Co., 1957.

———. *Galileo Galilei Sidereus Nuncius. Traduzione con Testo a Fronte e Note di Maria Timpanaro Cardini*. Florence: Sansoni, 1948.

———. *Galileo Galilei Sidereus Nuncius Nachricht von neuen Sternen. Dialog über die Weltsysteme (Auswahl). Vermessung der Hölle Dantes. Marginalien zu Tasso. Herausgegeben und eingeleitet von Hans Blumenberg.* Frankfurt am Main: Insel Verlag, 1965.

———. *El Mensajero de los Astros.* Translated by José Fernandes Chitt. Introduction by José Babini. Buenos Aires: Editorial Universitaria de Buenos Aires, 1964.

———. *Le messager céleste.* Translated by Alexandre Tinelis, Abbé de Castelet. Paris, 1681.

———. *Le Opere di Galileo Galilei.* Edizione Nazionale. 20 vols. Edited by Antonio Favaro. Florence: G. Barbera, 1890–1909; reprinted 1929–39, 1964–66.

———. *The Sidereal Messenger of Galileo Galilei and a Part of the Preface to Kepler's Dioptrics containing the original account of Galileo's astronomical discoveries. A translation with introduction and notes by Edward Stafford Carlos.* London, 1880; London: Dawson's of Pall Mall, 1960.

———. *Sidereus Nuncius; le message céleste. Texte établi et trad. par Émile Namer.* Paris: Gauthier-Villars, 1964.

———. *Sidereus Nuncius Magna, Longeque Admirabilia Spectacula pandens, suspiciendaque proponens unicuique, praesertim vero Philosophis, atque Astronomis, quae a Galileo Galileo Patritio Florentino Patavini Gymnasij Publico Mathematico Perspicilli Nuper a se reperti beneficio sunt observata in Lunae Facie, Fixis Innumeris, Lacteo Circulo, Stellis Nebulosis, Apprime vero in Quatuor Planetis Circa Iovis Stellam disparibus intervallis, atque periodis, celeritate mirabili circumvolutis; quos, nemini in hanc usque diem cognitos, novissime e Author depraehendit primus; atque Medicea Sidera Nuncupandos Decrevit.* Venice, 1610.

Geyl, Pieter. *The Netherlands in the Seventeenth Century. Part I, 1609–1648.* London: Ernest Benn, 1961.

Gingerich, Owen. "Discovery of the Satellites of Mars." *Vistas in Astronomy* 22 (1978): 127–32. Reprinted in *The Great Copernicus Chase.* Edited by Owen Gingerich. Cambridge: Cambridge University Press, 1988.

———. "Dissertatio cum Professore Righini at Sidereo Nuncio." In *Reason, Experiment, and Mysticism in the Scientific Revolution,* edited by Maria Luisa Righini Bonelli and William R. Shea, 77–88. New York: Science History Publications, 1975.

———. "The Mysterious Nebulae, 1610–1924." *Journal of the Royal Astronomical Society of Canada* 81 (1987): 113–27.

———. "Phases of Venus in 1610." *Journal for the History of Astronomy* 15 (1984): 209–10.

———. "The Satellites of Mars: Prediction and Discovery." *Journal for the History of Astronomy* 1 (1970): 109–15.

Grendler, Paul F. "The Roman Inquisition and the Venetian Press, 1540–1605." *Journal of Modern History* 47 (1975): 48–65. Reprinted in *Culture and Censorship in Late Renaissance Italy and France*. London: Variorum Reprints, 1981.

Gundel, Hans Georg, and Wilhelm Gundel. *Astrologumena: Die Astrologische Literatur in der Antike unde ihre Geschichte*. Beiheft 6. *Sudhoffs Archiv*. Wiesbaden: Franz Steiner, 1966.

Gundel, Wilhelm, and Hans Georg Gundel. *Astrologumena: Die Astrologische Literatur in der Antike unde ihre Geschichte*. Beiheft 6. *Sudhoffs Archiv*. Wiesbaden: Franz Steiner, 1966.

Hale, J. R. *Florence and the Medici: The Pattern of Control*. London: Thames & Hudson, 1977.

Handelman, George H., and Jane F. Koretz. "How the Human Eye Focuses." *Scientific American* 259, no. 1 (July 1988): 92–99.

Harriot, Thomas. Harriot Papers. West Sussex Record Office, Chichester, West Sussex. Petworth MSS HMC 241/4.

Harrison, Thomas G. "The Orion Nebula: Where in History Is It?" *Quarterly Journal of the Royal Astronomical Society* 25 (1984): 65–79.

Horky, Martinus. *Brevissima peregrinatio contra nuncium sidereum nuper ad omnes philosophos et mathematicos emissum a Galileo Galilaeo*. Bologna, 1610. In *Le Opere di Galileo Galileo,* Edizione Nazionale, edited by Antonio Favaro, 3:129–45. Florence, 1890–1909; reprinted 1929–39, 1964–66.

Hossenfelder, Malte. See Hans Blumenberg.

Hughes, David W. "Was Galileo 2,000 Years Too Late?" *Nature* 296 (18 March 1982): 199.

Humbert, Pierre. *Un amateur: Peiresc, 1580–1637*. Paris: Desclée, de Brouwer et cie., 1933.

———. "Joseph Gaultier de La Valette, astronome provençal (1564–1647)." *Revue d'histoire des sciences et de leurs applications* 1 (1948): 314–22.

Jaki, Stanley L. *The Milky Way: An Elusive Road to Science*. New York: Science History Publications; Newton Abbot: David & Charles, 1973.

Ilardi, Vincent. "Eyeglasses and Concave Lenses in Fifteenth-Century Florence: New Documents." *Renaissance Quarterly* 29 (1976): 341–60.

Kepler, Johannes. *Joannis Kepler astronomi opera omnia*. Edited by C. Frisch. 8 vols. Frankfurt and Erlangen, 1858–71.

———. *Johannes Kepler Gesammelte Werke*. Vols. 1–10, 13–19. Munich: C. H. Beck, 1937–.

———. *Kepler's Conversation with Galileo's Sidereal Messenger*. Translated by Edward Rosen. New York: Johnson Reprint Corp., 1965.

———. *Kepler's Dream*. Translated by John Lear. Berkeley: University of California Press, 1965.

———. *Kepler's Somnium*. Translated by Edward Rosen. Madison: University of Wisconsin Press, 1967.

Klug, Joseph. "Simon Marius aus Gunzenhausen und Galileo Galilei." *Abhandlungen der II. Klasse der Königlichen Akademie der Wissenschaften* 22 (1906): 385–526.

Koretz, Jane F., and George H. Handelman. "How the Human Eye Focuses." *Scientific American* 259, no. 1 (July 1988): 92–99.

Koyré, Alexandre. *From the Closed World to the Infinite Universe*. Baltimore: Johns Hopkins University Press, 1957; New York: Harper & Row, 1958.

Lear, John. *Kepler's Dream*. Berkeley: University of California Press, 1965.

Leonardo da Vinci. *Codex Leicester-Hammer*. In Jane Roberts, *Le Codex Hammer de Léonard de Vinci, les eaux, la terre, l'univers*. Paris: Jacquemart-André, 1982.

Marius, Simon. *Mundus Iovialis*. Nuremberg, 1614. Translated by A. O. Prickard. In "The Mundus Jovialis of Simon Marius." *Observatory* 39 (1916): 367–81, 403–12, 443–52, 498–503.

Meeus, Jean. "Galileo's First Records of Jupiter's Satellites." *Sky and Telescope* 24 (1962): 137–39.

Namer, Emile. *Sidereus Nuncius; le message céleste. Texte établi et trad. par Émile Namer*. Paris: Gauthier-Villars, 1964.

North, John D. "Thomas Harriot and the First Telescopic Observations of Sunspots." In *Thomas Harriot; Renaissance Scientist*, edited by John W. Shirley, 129–65. Oxford: Clarendon Press, 1974.

O'Malley, C. D., and Stillman Drake, eds., trans. *The Controversy on the Comets of 1618*. Philadelphia: University of Pennsylvania Press, 1960.

Oudemans, J. A. C., and J. Bosscha. "Galilée et Marius." *Archives Néerlandaises des Sciences Exactes et Naturelles*, series 2, 8 (1903): 115–89.

Pedersen, Olaf. "Sagredo's Optical Researches." *Centaurus* 13 (1968): 139–50.

Peters, William T. "The Appearances of Venus and Mars in 1610." *Journal for the History of Astronomy* 15 (1984): 211–14.

Peurbach, Johannes. *Theoricae Novae Planetarum.* Edited by Erasmus Reinhold. Wittenberg, 1553.

Propertius, Sextus. *The Elegies of Propertius.* Translated by E. H. W. Meyerstein. London: Oxford University Press, 1935.

Ptolemy. *Ptolemy's Almagest.* Translated by G. J. Toomer. London: Duckworth, 1984.

Righini, Guglielmo. *Contributo alla Interpretazione Scientifica dell'Opera Astronomica di Galileo.* Monograph 2, *Annali dell'Istituto e Museo di Storia della Scienza.* Florence, 1978.

———. "New Light on Galileo's Lunar Observations." In *Reason, Experiment, and Mysticism in the Scientific Revolution,* edited by Maria Luisa Righini Bonelli and William R. Shea, 59–76. New York: Science History Publications, 1975.

Risner, Friedrich. *Opticae Thesaurus.* Basel, 1572; New York: Johnson Reprint Corp., 1972.

Roberts, Jane. *Le Codex Hammer de Léonard de Vinci, les eaux, la terre, l'univers.* Paris: Jacquemart-André, 1982.

Robinson, Wade L. "Galileo on the Moons of Jupiter." *Annals of Science* 31 (1974): 165–69.

Roche, John. "Harriot, Galileo, and Jupiter's Satellites." *Archives internationales d'histoire des sciences* 32 (1982): 9–51.

Roffeni, Antonio. *Epistola apologetica contra caecam peregrinationem cuiusdam furiosi Martini, cognomine Horkij editam adversus nuntium sidereum.* Bologna, 1610. In *Le Opere di Galileo Galileo,* Edizione Nazionale, edited by Antonio Favaro, 3:191–200. Florence, 1890–1909; reprinted 1929–39, 1964–66.

Rosen, Edward. "The Authenticity of Galileo's Letter to Landucci." *Modern Language Quarterly* 12 (1951): 473–86.

———. "Did Galileo Claim He Invented the Telescope?" *Proceedings of the American Philosophical Society* 98 (1954): 304–12.

———. "Galileo on the Distance between the Earth and the Moon." *Isis* 43 (1952): 344–48.

———. "The Invention of Eyeglasses." *Journal for the History of Medicine and Allied Sciences* 11 (1956): 13–46, 183–218.

______. *Kepler's Conversation with Galileo's Sidereal Messenger.* New York: Johnson Reprint Corp., 1965.

______. *Kepler's Somnium.* Madison: University of Wisconsin Press, 1967.

______. *The Naming of the Telescope.* New York: Henry Schuman, 1947.

______. "Stillman Drake's *Discoveries and Opinions of Galileo.*" *Journal for the History of Ideas* 48 (1957): 439–48.

______. "The Title of Galileo's *Sidereus Nuncius.*" *Isis* 41 (1950): 287–89.

Sarpi, Paolo. *Fra Paolo Sarpi, Lettere ai Protestanti.* Edited by Manlio Duilio Busnelli. 2 vols. Bari: Gius. Laterza & Figli, 1931.

Schevill, Ferdinand. *The Medici.* New York: Harcourt, Brace & Co., 1949; New York: Harper, 1960.

Shea, William, and Maria Luisa Righini Bonelli, eds. *Reason, Experiment and Mysticism in the Scientific Revolution.* New York: Science History Publications, 1975.

The Sidereal Messenger. Edited by W. W. Payne. Vols. 1–10. Northfield, Minnesota, 1882–91. Vols. 11–13 (1892–94) are entitled *Astronomy and Astro-Physics.* Edited by W. W. Payne and G. E. Hale. Superseded by *Astrophysical Journal.*

The Sidereal Messenger, a monthly journal devoted to astronomical science. Edited by O. M. Mitchel. Vols. 1–2, vol. 3, nos. 1–2. Cincinnati, 1846–48.

Sirturi, Girolamo. *Telscopium: Sive ars perficiendi.* Frankfurt, 1618.

Suetonius. *The History of Twelve Caesars translated into English by Philemon Holland anno 1606.* 2 vols. London: David Nutt, 1899.

Swift, Jonathan. *Gulliver's Travels.* In *The Prose Works of Jonathan Swift,* 14 vols. edited by Herbert Davis, vol. 13. Oxford: Basil Blackwell, 1939–68.

Tinelis, Alexandre, Abbé de Castelet. *Le messager céleste.* Paris, 1681.

Tuckerman, Briant. *Planetary, Lunar and Solar Positions A.D. 2 to A.D. 1649 at Five-Day and Ten-Day Intervals.* American Philosophical Society, *Memoirs* 59 (1964).

Van Helden, Albert. " 'Annulo Cingitur': The Solution of the Problem of Saturn." *Journal for the History of Astronomy* 5 (1974): 155–74.

______. *The Invention of the Telescope.* American Philosophical Society, *Transactions* 67, part 4 (1977).

______. *Measuring the Universe: Cosmic Dimensions from Aristarchus to Halley.* Chicago: University of Chicago Press, 1985.

______. "Saturn and His Anses." *Journal for the History of Astronomy* 5 (1974): 105–21.

Vitello. *Perspectiva*. In *Opticae Thesaurus*. Edited by Friedrich Risner. Basel 1572; New York: Johnson Reprint Corp., 1972.

Westfall, Richard S. "Science and Patronage: Galileo and the Telescope." *Isis* 76 (1985): 11–30.

Westman, Robert S. "The Astronomer's Role in the Sixteenth Century: A Preliminary Study." *History of Science* 18 (1980): 105–47.

Whitaker, Ewen. "Galileo's Lunar Observations and the Dating of the Composition of 'Sidereus Nuncius.' " *Journal for the History of Astronomy* 9 (1978): 155–69.

Witelo. *Optica* (*Perspectiva*). In *Opticae Thesaurus*. Edited by Friedrich Risner. Basel, 1572; New York: Johnson Reprint Corp., 1972.

Wodderburn, John. *Quatuor Problematum quae Martinus Horky contra Nuntium Sidereum de quatuor planetis novis disputanda proposuit*. Padua 1610. In *Le Opere di Galileo Galilei*, Edizione Nazionale, edited by Antonio Favaro, 3:147–78. Florence, 1890–1909; reprinted 1929–39, 1964–66.

Xi Ze-zong. "The Sighting of Jupiter's Satellite by Gan De 2000 Years before Galileo." *Chinese Astronomy and Astrophysics* 5 (1981): 242–43.

INDEX